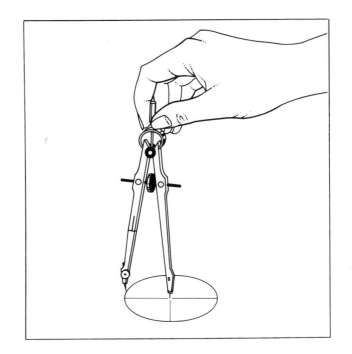

THE TECHNOLOGY OF DRAFTING

With an Introduction to Computer-Aided Drawing

EDWARD A. MARUGGI

Rochester Institute of Technology

Technical illustrations by Albert F. Luiz

Merrill Publishing Company
A Bell & Howell Information Company
Columbus Toronto London Melbourne

3-94

19824176

Published by
Merrill Publishing Company
A Bell & Howell Information Company
Columbus, Ohio 43216

This book was set in Univers and Rockwell.

Administrative Editor: John Yarley
Production Coordinator: JoEllen Gohr
Art Coordinator: Ruth Kimpel
Cover Designer: Brian Deep
Cover Photo: Larry Hamill

Library of Congress Catalog Card Number: 88-63912
International Standard Book Number: 0-675-20762-2
Printed in the United States of America
1 2 3 4 5 6 7 8 9—92 91 90 89

To my son, Stephen.

MERRILL SERIES IN MECHANICAL, INDUSTRIAL, AND CIVIL TECHNOLOGY

Preface

The Technology of Drafting grew out of my several years of successfully using its material to instruct students in degree-level programs at Rochester Institute of Technology, in New York. It describes current methods for producing various types of technical drawings required by industry and provides drafting technology fundamentals that can be easily applied to computer-aided drafting (CAD) systems. The presentation has been developed to accommodate students at various levels of experience. It is appropriate for curricula in high schools, community colleges, technical institutes, vocational programs, industrial training programs, and continuing education courses, and it is relevant for programs in drafting, machine and mechanical technology, industrial technology, engineering technology, electromechanical technology, computer technology, quality control, and inspection—in short, any program that requires a basic knowledge of mechanical drafting techniques related to the technological aspects of industry.

The organization of the text allows the individual learner to progress from simple to complex work. Each section is prefaced by clearly stated learner outcomes, and performance is measured against these expectations by review exercises at the end of each section and by the practice exercises in the accompanying workbook, Workbook in the Technology of Drafting. The text is sufficiently illustrated to permit its use in a "self-paced" mode, or the instructor can choose to lecture on each individual topic.

The Technology of Drafting is designed to reflect the format of a Standard Drafting Practice (SDP) handbook or a Drawing Requirements Manual (DRM), both of which are widely used by industry in an effort to standardize drafting procedures. For the most part, the drawing requirements are based on practices of the U.S. Department of Defense (DOD), the American National Standards Institute (ANSI), and other engineering and professional groups. The section and subsection designations will familiarize the students with the style of an SDP handbook or a DRM prior to entering the world of work.

The text is self-contained; that is, no additional materials are needed. The many tables and charts in Appendixes A through N contain all the reference material the student will need to successfully complete the coursework. Following the appendixes is a glossary containing concise definitions of all important terms.

I would like to thank the various corporations, organizations, and government agencies that contributed their materials for use in the text. They include the American National Standards Institute; Apple Computer, Inc.; Autodesk, Inc.; CADAM, Inc.; Chartpak/Pickett; Hewlett-Packard Corp.; Houston

Instruments; Intergraph Corp.; International Business Machines Corp.; Olson Manufacturing Inc.; Paper Machinery Corp.; J.S. Staedtler, Inc.; Summagraphics Corp.; T & W Systems, Inc.; and Rotor Clip Inc. I would also like to acknowledge the technical assistance and critique of Eder Benati, NTID at RIT, Rochester, New York; Anthony Scalise, Harris Corp., Rochester, New York; Bruce Henry, Eastman Kodak Co., Rochester, New York; and Albert Luiz, Indian Harbour Beach, Florida.

In addition, special thanks go to Albert Luiz for producing the technical illustrations and to the following professionals in the media production department of the National Technical Institute for the Deaf at RIT for their cooperation: Willard Yates, Sarah Perkins, Jorge Samper, and Mark Benjamin. Finally, I am also grateful to the following prepublication reviewers for their helpful comments and suggestions: Bill Shorthill and James Loew, Stark Technical College; George Baggs, Augusta Technical Institute; Madison Ashburn, Kansas Technical Institute; Kurt Commons, City College of San Francisco; and Russ Schultz, Hawkeye Institute of Technology.

Contents

Section 1

Introduction to the Technology of Drafting

LEARNER OUTCOMES

The student will be able to:

- Define the technology of drafting

- State the purpose of the language of drafting technology

- Identify 12 common technical drawings required by industry

- List two organizations responsible for standardizing drafting practices

- Identify the father of descriptive geometry and of projection principles

- List the equipment necessary to perform manual drafting in the modern drafting laboratory

- Identify the elements of a computer-aided drafting workstation

- Demonstrate learning through the successful completion of review and practice exercises

1.1 DEFINITION

Drafting technology is a method of communication and is the language of industry. As distinguished from drawing as a fine art practiced by artists in pictorial representation, drafting technology is a descriptive form of communication, whereas drawing as a fine art is an aesthetic expression. The term **technology of drafting** is not new. **Technology** has been defined as "a practical application of a branch of knowledge that deals with industrial arts, applied science, or engineering," and **drafting** has been defined as "the systematic representation and dimensional specification of mechanical and architectural structures including drawing, sketching, and designing." So, for our purpose we may conclude that the technology of drafting is the art of mechanical drawing as it pertains to the field of engineering. In communication through the means of drafting technology, lines, symbols, shapes, views, dimensions, and detailed information take the place of many words. This graphic technical language is used in engineering departments and manufacturing facilities, between contractors and customers, and between buyers and vendors (see Figure 1.1).

Designing a product from initial concept to final assembly requires the energies and talents of various technical and professional people. One of the key people in this process is the drafter, who takes a design and, through skill in drafting technology and knowledge of its practices and procedures, completes the task so that parts, subassemblies, and assemblies can be produced. Figure 1.2 represents today's drafter at work at manual drafting and computer-aided drafting (CAD) workstations.

The industrial sector requires many types of drawings and diagrams for the various applications and uses of consumer, commercial, and military products that are produced throughout the world.

Some types of drawings are required more often than others, depending on the nature of the industry's product or product line, and on the structure and size of the company. Since the primary function of this book is to serve as a mechanical text, the following list identifies the more common types of technical drawings and diagrams required by industry today:

- The assembly drawing
- The control drawing
- The design layout drawing
- The detail drawing
- The erection drawing
- The expanded assembly drawing, also referred to as the "exploded view" drawing
- The installation drawing
- The kit drawing
- The mechanical schematic diagram
- The modification drawing
- The numerical control (N/C) drawing
- The packaging drawing
- The piping diagram
- The proposal drawing
- The tabulated assembly drawing
- The tube bend drawing

The practices described in this textbook are those that have become standardized over the years through the efforts of the United States Department of Defense (DOD), the American National Standards Institute (ANSI), and other fine professional and educational societies and organizations. These practices are accepted by industrial contractors of military products as well as by large and medium-size companies producing consumer and commercial products.

1.2 PURPOSE

In today's marketplace, many American consumer products are manufactured in countries such as Japan, Taiwan, and Korea. Automakers from Japan and Germany are producing cars in the United States. An American military aircraft's basic airframe is manufactured in the United States, but hundreds of its assemblies, subassemblies, and parts are made in other nations around the free world. The universal language, the principles of which are essentially the same throughout the world and tie countries together, is the technology of drafting. The purpose of this language is to provide visually oriented data that is usable by technical, engineering, and manufacturing personnel to produce goods and services.

FIGURE 1.1
Two people communicating via technical graphics.

FIGURE 1.2
Drafters in the workplace.

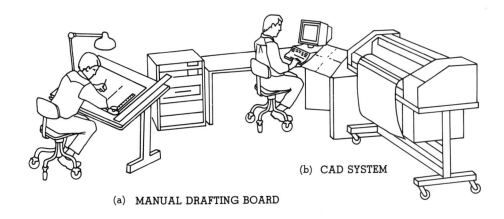

(b) CAD SYSTEM

(a) MANUAL DRAFTING BOARD

1.3 HISTORY

For thousands of years, various forms of drawings have been produced to illustrate designs, ideas, and concepts. They have been helpful in the construction of various structures since early times. Primitive humans made sketches and drawings with crude tools such as sharp rocks, sticks, and other available sharp instruments. A graphic language called hieroglyphics was used by the Egyptians, between 4000 and 3000 B.C., to portray and represent objects. The papyrus plant provided a parchment type of writing and drawing surface, like paper, that allowed more advanced and elaborate sketches to be produced. An example of this Egyptian language is shown in Figure 1.3.

The Romans, early in the Christian era, were brilliant architects who were very skillful in producing technical graphics of buildings that were being constructed. They used crude variations of today's drafting tools such as pens, compasses, and straightedges to develop floor and elevation plans of structures. They also produced a type of drawing called the **perspective**, which shows a structure as the eye sees it.

At about the middle of the 15th century, Leonardo da Vinci, the great Italian artist, sculptor, engineer, and scientist, was recording his thoughts and concepts for a great number of mechanical designs and inventions. Da Vinci's primary mode of expression was the **pictorial drawing** (Figure 1.4).

The father of descriptive geometry was a Frenchman, Gaspard Mongé, who lived in the 18th and 19th centuries. His book *La Géometric Déscriptive* was written and published while he was a professor of mathematics at the Polytechnic University in France. He is credited with developing the principles of projections which are used today and which form the basis for technical graphics. Since that time, technical graphics has become the principal vehicle for the communication of technology and is now an integral part of technical education and engineering.

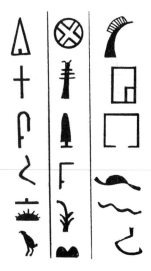

FIGURE 1.3
Egyptian hieroglyphics.

FIGURE 1.4
Da Vinci pictorial.

1.4 CURRENT PRACTICES

This is a time of rapid change in the field of drafting technology. The era of the four-legged drafting board (see Figure 1.5), with a straightedge, a high stool, an eye shade, and a single dangling light bulb to illuminate both board and drawing, has all but disappeared. The emphasis on technical drafting to convey an idea, concept, or thought is very much with us, but the tools and equipment have changed, and so has the technical education and training of the designer and drafter.

The technology of drafting can be performed with a minimal investment in equipment and basic tools, or one can spend large sums of money in outfitting a drafting laboratory. Modern drafting rooms are equipped with large, full-size manual drafting stations that include drafting boards and drafting machines designed to move in any direction quickly and accurately to accommodate the technical needs of the drafter. In addition, drafting chairs are comfortable, supporting, and adjustable to discrete heights. The manual drafting station usually includes a reference table located behind or to the side of the draft-

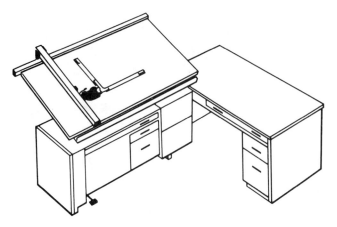

FIGURE 1.6
Modern manual drafting station.

er. This table is used for reference materials the drafter needs, which may include drawings, layouts, drafting manuals, reference books, and catalogs. Figure 1.6 illustrates a typical manual drafting station used in a modern drafting room.

The automated drafting station is called the computer-aided graphics workstation. It is used for producing computer-aided drafting (CAD) data and consists of two basic elements: **hardware** and **software**. The system illustrated in Figure 1.7 is a complete **interactive** computer system for two-dimensional or three-dimensional drafting. In-depth coverage of CAD is presented in Section 16.

Current practices in drafting can be found in "Dimensioning and Tolerancing for Engineering Drawings," a specification of the American National Standards Institute (ANSI), revised in 1982, called ANSI Y14.5M. This document establishes and illustrates current methods to be used in specifying dimensions and tolerances on drawings. Also, there appears to be a new awareness of, and a need to move more directly and rapidly to, the metric system of measurement. Major multinational industries have converted to **SI units** (Système Internationale d'Unités) in an effort to support their worldwide operations in dealing with countries committed to the metric system.

1.5 SUMMARY

The technology of drafting is a descriptive form of communication used around the world by technical and engineering personnel to develop, design, and draw parts, subassemblies, assemblies, and systems for consumer, commercial, and military products.

FIGURE 1.5
A drafter's workstation of the past.

FIGURE 1.7
Computer-aided graphics
workstation.

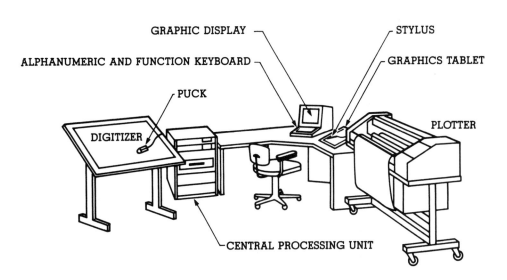

Several types of technical drawings required by industry were identified. A brief history of the technology of drafting was traced from primitive times to the present. Current practices were outlined and equipment required for both manual drafting and computer-aided drafting (CAD) were listed and defined.

1.6 REVIEW EXERCISES

1. What is the technology of drafting?

2. List 12 types of technical drawings and diagrams used in industry.

3. Name the graphics language used by the Egyptians between 4000 and 3000 B.C.

4. Who is considered the father of descriptive geometry and is credited with developing the principles of projection?

5. In the 15th century, who used the pictorial drawing as his mode of expression for a great number of mechanical designs and inventions?

6. List four pieces of equipment used in the modern manual drafting room.

7. What is a reference table used for in a drafting room?

8. What are two basic elements of a computer-aided graphics workstation?

9. List six pieces of hardware used in a computer-aided graphics workstation.

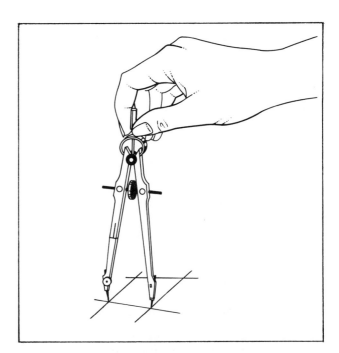

Section **2**

Use of Drafting Equipment and Tools

LEARNER OUTCOMES

The student will be able to:

- State the purposes of drafting equipment and tools

- Name four input devices and two output devices that are included in the automated drafting laboratory

- Explain the diazo process of reproducing drawings

- Identify seven manual tools used by the drafter

- Name three types of media used to produce manual graphics

- List three types of measurement systems

- Demonstrate learning through the successful completion of review and practice exercises

2.1 PURPOSE

Drafting equipment and **tools** aid the drafter in producing accurate, clear, and legible technical information in a timely manner to meet predetermined production schedules. There are two types of equipment and tools used in the modern drafting laboratory: manual and automated. Time is money, and if a drafter can produce quality technical graphics quickly and efficiently, he or she will be a tremendous asset to the organization.

2.2 EQUIPMENT

2.2.1 Manual Equipment

The **manual drafting laboratory** may be equipped with several different types of drafting furniture. The **drafting board** may take the form of a completely movable graphics surface that can be adjusted to any desired angle or height. Adjustment can be made with the foot or the hand, or both. An example is shown in Figure 2.1.

The **drafting machine**, which is normally mounted on the top edge of the drafting board, is available to accommodate both left-handed and right-handed drafters. Because of their mobility, these machines allow rapid and accurate drawing of horizontal, vertical, and angled lines with the flip of a lever or the push of a button. Figure 2.2 illustrates a drafting machine mounted on a drafting board.

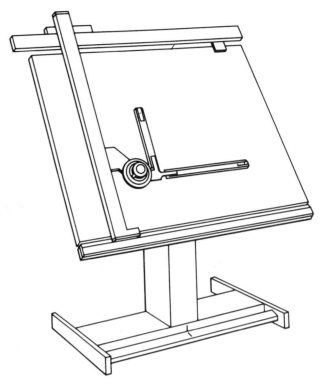

FIGURE 2.2
Drafting machine mounted on a board.

For technical or financial reasons, a table-top drafting board is sometimes used. It is utilized with either a **T-square** or a **parallel straightedge** to produce technical drawings. An example of a table-top

FIGURE 2.1
Movable graphics surface.

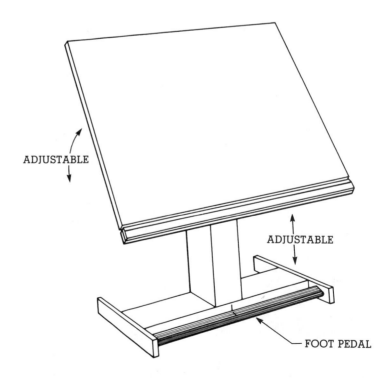

ADJUSTABLE

ADJUSTABLE

FOOT PEDAL

FIGURE 2.3
Table-top drafting board.

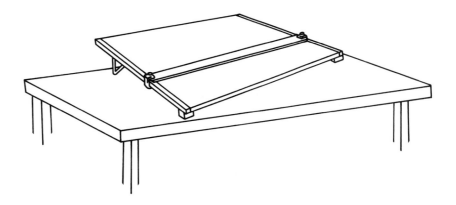

drafting board is illustrated in Figure 2.3. T-squares and straightedges can be used only to draw horizontal lines. Triangles must be used for drawing vertical or angled lines.

Figure 2.4 shows how the parallel straightedge and the T-square are used for drawing horizontal lines.

A **file** is used for permanent storage of technical drawings, drawings and layouts in process, or print paper. Examples of two styles of drawing files are shown in Figure 2.5.

Special tables, sometimes including drawers, are used by drafters (see Figure 2.6). They are usually located to the side of the manual drafting board or behind the drafter. Such a table is called a **reference table** and provides a surface for the drafter to place drawings, prints, drafting manuals, and reference books. The drawers may be used as a file for vendor catalogs, project files, technical data, and the like.

The **diazo system** of reproducing engineering drawings has been the primary method of making prints since the 1950's. The dry diazo process is based on the reaction of light-sensitive paper to ultraviolet light. When exposed to ultraviolet light, this paper has the characteristic of decomposing into a colorless substance. Ammonia is usually employed as the developing liquid in this process, and its odor can be a problem if the equipment is not vented properly. This type of printing machine normally produces **blueline prints**, incorrectly referred to as "blueprints." The diazo process also produces blackline, brownline, and sepia prints on a wide range of polyester films, papers, and other base materials. Diazo machines are available in many sizes with various accessories. These printers are fast, flexible, and economical. The model and style required usually depend on the reproduction needs of the organization. Figure 2.7 illustrates one blueline diazo process. In addition, copier manufacturers are producing a variety of electrostatic printers that offer both enlarging and reduction capabilities.

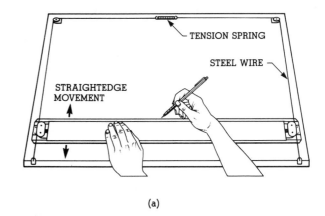

(a)

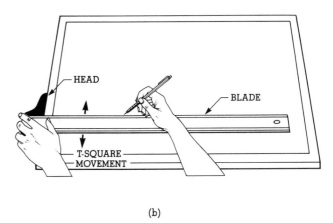

(b)

FIGURE 2.4
(a) Parallel straightedge; (b) T-Square.

2.2.2 Automated Equipment

Automated drafting laboratory equipment is significantly different from that used in the manual drafting area. In today's medium-to-large-size companies, 25 to 90 percent of all mechanical and electronic technical graphics are produced on some form of automated drafting system. Radical changes have taken place in the method of technical communication. **Automated drafting** is accomplished by use of an integrated combination of items called hardware and software. Not too long ago, very few people

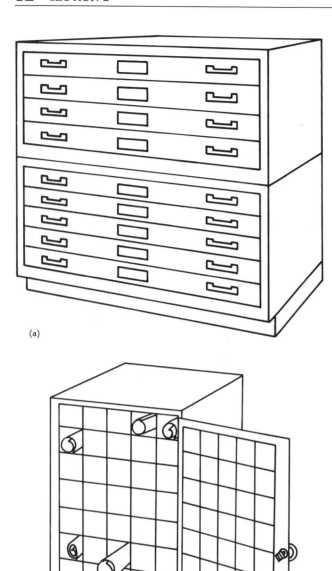

(a)

(b)

FIGURE 2.5
(a) Flat file; (b) Pigeonhole file.

FIGURE 2.6
Reference table.

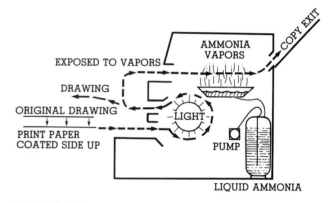

FIGURE 2.7
Blueline diazo process.

knew the meanings of terms such as **hardware**, **software**, **CRT**, **operating system**, **digitizer**, and **central processing unit (CPU)**. The modern drafting laboratory is made up of input and output devices, and the drafter's title has changed to **operator**, **technician**, or **technologist** of computer-aided drafting (CAD). Technical drawings are being produced electronically through digitally controlled machines. Input devices include hardware items with names such

as alphanumeric keyboard, digitizer, light pen, mouse, touch pen, function keyboard, puck, stylus, and graphics tablet. Output devices include monitors, printers, and plotters. A representative example of a state-of-the-art automated drafting system is illustrated in Figure 2.8.

The automated drafting laboratory and, specifically, computer-aided drafting will be covered in more depth in Section 16.

2.3 TOOLS

For manual production of mechanical-type technical drawings, several special tools of the trade are required. These tools are important aids in making certain that clear, detailed, and concise graphics meeting acceptable industrial standards are produced in a minimum amount of time.

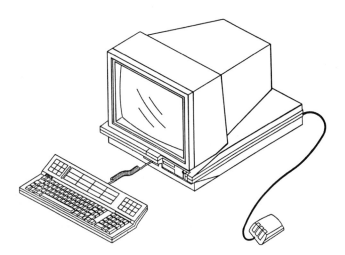

FIGURE 2.8
Automated drafting system.

2.3.1 Leads, Lead Holders, and Sharpeners

Leads and mechanical lead holders have, for the most part, replaced the pencil for the drawing of technical graphics. Several different grades of **lead** are available, ranging from very hard to very soft. Table 2.1 identifies the most popular leads and their primary uses. Leads are graded so that the higher the "H" number, the harder the lead. Table 2.1 should be used as a guide. Drafters have personal preferences as to which grade of lead produces the desired quality. In addition, the grade of lead used is often determined by the type of drawing medium employed, the weight of hand, and the weights and kinds of lines required. Two types of leads are used for manual drafting: (1) graphite lead, which is used on all paper surfaces including vellum; and (2) plastic lead, which normally is restricted to use with a polyester film material called Mylar (a registered product of the DuPont Company).

Leads cannot be used without a **mechanical lead holder**. Lead holders, or mechanical pencils as they are sometimes called, are available in many types and vary according to lead diameter and type

TABLE 2.1
Leads and their usage.

LEAD HARDNESS	USAGE
6H 5H 4H 3H 2H H	GUIDE LINES, CONSTRUCTION LINES CENTER LINES MEDIUM WEIGHT USES AND FOR GENERAL DRAWING THICK LINES, LETTERING SKETCHING

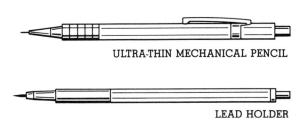

ULTRA-THIN MECHANICAL PENCIL

LEAD HOLDER

FIGURE 2.9
Lead holders.

of lead. Figure 2.9 shows two of the most common types of holders.

The lead sharpener used by drafters is called a **lead pointer**. It is so called because it does just that—it produces points on leads used in lead holders. The pointer produces a conical-shape point that allows the drafter to produce quality lines on drawings. The lead pointer is available in both manually operated and electrically operated varieties. Examples of each are illustrated in Figure 2.10.

(a)

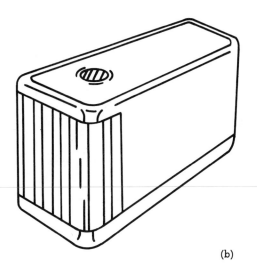

(b)

FIGURE 2.10
(a) Manually operated lead pointer; (b) Battery operated lead pointer.

2.3.2 The Triangle

The drafter normally uses three types of **triangles**: the 45-degree triangle, the 30/60-degree triangle, and the adjustable triangle, all of which are shown in Figure 2.11.

Triangles are manufactured in different sizes, are made of plastic, and are available in clear or colored material. Used individually, the 45-degree and 30/60-degree triangles allow one to draw lines at 30, 45, 60, and 90 degrees from the horizontal. When used together, however, they can produce additional angles of 15 and 75 degrees, as illustrated in Figure 2.12.

The adjustable triangle includes a protractor so that any angle in one-half-degree increments can be obtained between 0 and 90 degrees. An adjusting nut allows the drafter to firm up the desired angle before use.

Figure 2.13 shows the proper method for drawing angled lines using a triangle.

2.3.3 The Compass

The **compass** is used to draw circles and arcs. Compasses can be used for both ink and lead work and are available in various sizes and styles to meet different drawing requirements.

2.3.3.1 The Bow Compass

The center wheel adjustment **bow compass** is the compass most commonly used in drafting laboratories. Adjustment by means of the center wheel allows the drafter to set the particular arc or radius to be drawn and ensures that the setting will not move out

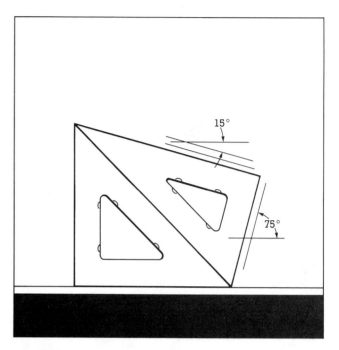

FIGURE 2.12
Drawing 15-degree and 75-degree angles using two triangles with a straightedge or T-square.

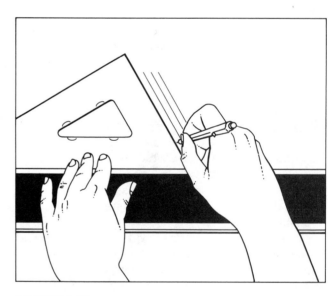

FIGURE 2.13
Proper method for drawing angled lines.

of adjustment until a readjustment is required. This type of compass and its use are demonstrated in Figure 2.14.

2.3.3.2 The Beam Compass

A tool not used very frequently is the **beam compass**. The purpose of the beam compass is to draw large circles and arcs that cannot be drawn with a bow

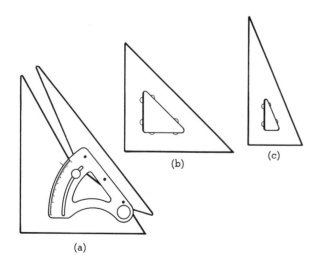

FIGURE 2.11
(a) Adjustable triangle; (b) 45-degree triangle; (c) 30/60-degree triangle.

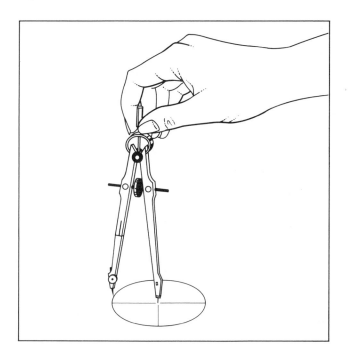

FIGURE 2.14
How to use a center wheel adjustment bow compass.

compass. This compass includes a bar and two movable point assemblies to secure lead, points, or a pen and must be positioned using both hands, as shown in Figure 2.15.

2.3.4 The Dividers

The **dividers** are similar in construction to a bow compass except that the lead end of the compass is replaced with an additional metal point, as pictured in Figure 2.16. Dividers are used for transferring measurements, dividing lines, and laying out series of equal distances required in geometric constructions, in descriptive geometry, or on development drawings.

2.3.5 The Template

Templates are widely used by drafting laboratory personnel. A **template** is a thin (usually 1/32 to 1/16 inch thick) plastic aid for producing technical drawings. It is a time-saving device that allows rapid and accurate drawing of shapes and forms which are difficult to produce by other methods or tools. Many types of templates are available for drawing circles, ellipses, electronic symbols, logic symbols, architectural symbols, screw threads, hardware, and specialty building trade items.

When using a flat template (without nibs or standoffs to keep it away from the drawing), one should be careful to lift the template after drawing each symbol or character. In order to keep the drawing surface clean, do not slide the template along the drawing surface. When inking, wait for the ink to dry before lifting or moving the template. Figure 2.17 illustrates how to use a template.

2.3.6 The Curve

Curves are used to draw irregular or noncircular shapes which are frequently required in producing development, ellipse-, parabola-, and hyperbola-type drawings. They are often referred to as French curves or irregular curves. Curves are available in several different shapes and sizes (see Figure 2.18) and are capable of producing an unlimited number of irregular shapes.

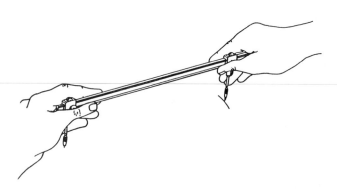

FIGURE 2.15
Producing a large arc using a beam compass.

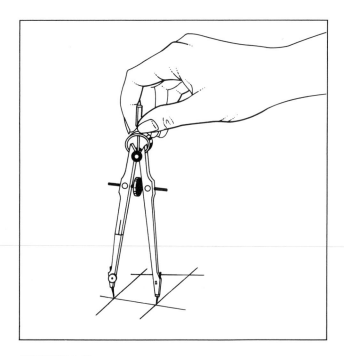

FIGURE 2.16
Use of dividers.

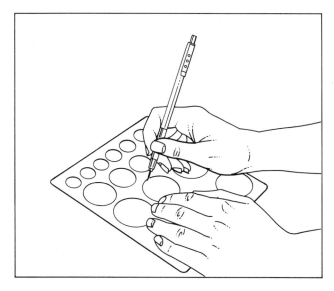

FIGURE 2.17
How to use a circle template.

FIGURE 2.18
Irregular curves.

Curves are used to produce curved lines which cannot be drawn with a compass or a template. Curves are thin plastic devices which are produced in clear or colored tones. The curve is normally used to draw a shape which is determined by a series of plotted point locations. The curve is used to connect the plotted points with a line, as shown in Figure 2.19.

2.3.7 The Eraser

The **eraser** is used for making corrections and changes in technical drawings. There are pencil-type erasers, block-type erasers, soft and hard erasers, white and pink erasers, and plastic erasers. It is common practice to use a soft eraser to eliminate unwanted pencil lines. Hard erasers seem to contain a lot of grit which may tend to cause severe damage to the surface of the drawing medium. Vinyl erasers are suggested for use on a polyester film base medium. Examples of commonly used erasers are illustrated in Figure 2.20.

The **electric eraser** provides a more rapid method for making corrections or changes. More

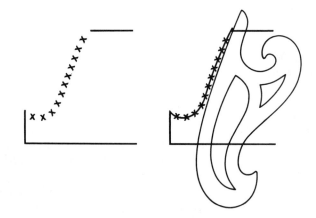

FIGURE 2.19
Curve drawn from plotted points.

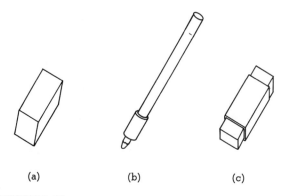

(a) (b) (c)

FIGURE 2.20
Commonly used erasers: (a) Pink pearl; (b) Pencil type; (c) Plastic.

and more drafters are using electric erasers, which are available in both cord and cordless types. The cordless type can be used anywhere, is not restricted to cord length, and operates on batteries which are rechargeable. The disadvantage of the electric eraser is that beginning drafters, if they are not careful, may have the tendency to abrade through the drawing medium. Figure 2.21 shows commonly used cord-type and cordless electric erasers.

2.3.7.1 The Erasing Shield

The **erasing shield** is an ultra-thin piece of metal, with cutouts of various shapes, that is used as an aid in erasing segments of lines, arcs, or circles. The shield is used by placing it over the line segment to be erased; it should be held firmly in place. Care should be taken not to erase too hard or a hole through the medium may result. The use of an erasing shield is demonstrated in Figure 2.22.

2.3.8 The Drawing Medium

Materials for producing manual technical drawings include paper, cloth, and film-base materials. Each

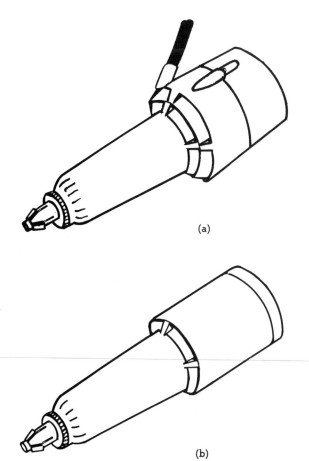

FIGURE 2.21
Electric erasers: (a) Cord type; (b) Cordless.

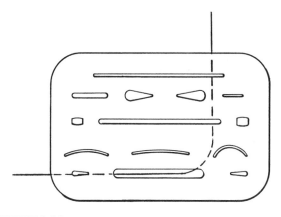

FIGURE 2.22
Use of an erasing shield.

of these materials is called a **drawing medium**. The two types of media most frequently used are (1) **vellum**, a white, translucent, rag-content paper, and (2) a film that is durable, is dimensionally stable, and can be stored for many years with no degradation of drawing quality. The film-type medium is available under several commercial trademark names and is really a **polyester film**. The advent of the office copier that uses a toner has made it possible for sketches and drawings measuring 8½ by 11 inches, 8½ by 14 inches, or larger to be reproduced on plain, lined, or unlined bond and on reproducible and nonreproducible grid paper. These copiers allow engineering personnel to provide drafters with reductions, enlargements, or copies of segments of drawings. Drawing media are available in standard sizes, either plain or with the company's format preprinted on one side. The practice exercises in the workbook represent a printed format on a white bond medium.

Table 2.2 identifies frequently used drawing sizes that have become standard and are in accordance with MIL-STD-100A, a military drawing standard for industries under contract to the U.S. Government. Roll-size drawing media are normally used for layouts, large assembly drawings, exploded-view drawings, or very large drawings, because of the complexity of the equipment involved.

2.3.9 Line Convention

Two weights, or widths, of lines are recommended by the American National Standards Institute specification ANSI Y14.2M-1979 for the common types of technical drawings required by industry. These weights are thin (.015 to .022 inch) and thick (.030 to .038 inch). All lines should be clean-cut, opaque, and of uniform width throughout their lengths. The various types of lines that are used are characterized as follows and are illustrated in Figure 2.23.

TABLE 2.2
Frequently used drawing sizes.

SIZE (IN INCHES)			LETTER SIZE
WIDTH		LENGTH	
11	x	8½	A
11	x	17	B
17	x	22	C
22	x	34	D
34	x	44	E

SIZE (MILLIMETERS)			LETTER SIZE
WIDTH		LENGTH	
210	x	297	A4
297	x	420	A3
420	x	594	A2
594	x	841	A1
841	x	1189	A0

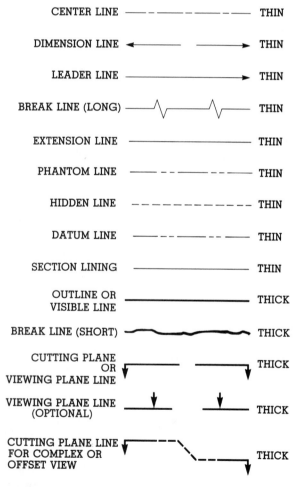

FIGURE 2.23
Line characteristics.

- **Center line** A thin line made up of long and short dashes, alternately spaced and consistent in length, beginning and ending with a long dash. Center lines cross each other without voids. Very short center lines may be unbroken if there is no possibility of confusion with other lines. (*Use:* To indicate the axis of a part or feature, a path of motion, or the theoretical line about which a part or feature is symmetrical.)
- **Dimension line** A thin line the same width as the center line, terminating in arrowheads at both ends and unbroken except where the dimension is placed. (*Use:* To indicate the extent and direction of a dimension.)
- **Leader line** A thin line the same width as the center line, terminating in an arrowhead or dot. An arrowhead should always terminate at a line. Generally, a leader line should be an inclined line except for a short horizontal portion extending to mid-height of the first or last letter or digit of the note. (*Use:* To indicate a part or portion of a drawing to which a number, note, or reference applies.)
- **Break line** A thin line with freehand zigzags for long breaks, or a thick freehand line for short breaks. (*Use:* To show an area or portion of a part that has been removed to show internal detail, to limit a partial section or view, or to eliminate repeated detail.)
- **Extension line** A thin, continuous line that does not touch the outline of the object. (*Use:* To extend points or planes to indicate dimensional limits.)
- **Phantom line** A thin line consisting of single long and double short dashes, alternately and evenly spaced, with a long dash at each end. (*Use:* To show alternate positions of parts, relative positions of adjacent parts shown for reference, or to eliminate repeated detail.)
- **Section lining** A thin, continuous line usually drawn at a 45-degree angle. (*Use:* To indicate surfaces exposed by a section cut.)
- **Hidden line** A thin line of short dashes (approximately 1/8 inch long), closely and evenly spaced. It is less prominent than an outline or a visible line. A hidden line begins and ends with dashes in contact with the lines that define its endpoints. (*Use:* To show hidden features of a part.)
- **Outline** or **visible line** A thick line, usually the most prominent one on a drawing. (*Use:* For all lines on the drawing representing visible edges or lines of an object.)
- **Cutting plane** or **viewing plane line** A thick line ending with arrowheads drawn at 90 degrees to the cutting or viewing plane line, to indicate the viewing direction. (*Use:* A cutting plane line indicates the plane in which a section is taken; a view-

ing plane line indicates the plane from which a surface is viewed.)

Examples of line convention appear in Figure 2.24.

2.4 MEASUREMENT

Measurement may be defined as the determination or estimation of a value, dimension, or quantity by comparison with some known standard. There are several systems of measurement that are used throughout the world, including the metric system, the inch-foot system, the dual-dimensioning system, and the redefined metric (SI) system. Because so many United States corporations are multinational, which means they operate internationally, engineers, drafters, and manufacturing personnel must be prepared to exchange technical information internationally. At times, this is difficult because of the different systems that are used for measurement. In the United States there has been a strong tendency to retain the inch-foot (English) system of measurement while most other countries utilize the metric system. Since the metric system is not universally accepted in the United States, it is important for producers of drafting technology to be knowledgeable about all the major types of measurement systems.

2.4.1 The Scale

One important basic skill for the drafter is the ability to measure using **scales**. A scale is a device made of wood, plastic, or metal that is used for determining measurements as well as for showing parts at a reduced or enlarged, suitable size.

Scales are available in two basic shapes: flat and triangular. The commonly used scales are shown in Figure 2.25. The triangular scale has an advantage over the flat scale in that it has six surfaces on which different scales appear whereas the flat scale has four at most. Most drafters prefer the flat scale because it is easier to use. At times, the beginning drafter will have difficulty in reading the triangular scale because reading from both ends may be required.

2.4.1.1 The Inch-Foot Scale

An example of the **inch-foot scale** is the mechanical engineer's scale. It is available in both fractional and decimal forms. This type of scale is usually divided into the common fractions of an inch (1/4, 1/8, 1/16, and 1/32). An example of such a scale is shown in Figure 2.26.

The **decimal scale**, which is widely used in industry, especially for electronics and mechanical-type technical drafting, is divided primarily into fiftieths of an inch, with each unit being equal to .02 inch. This is especially suitable for use with two-place decimal systems. An illustration of an engineer's decimal scale is shown in Figure 2.27.

The commonly used fractional and decimal divisions used for mechanical engineers' scales are identified in Tables 2.3 and 2.4.

On drawings of machine parts it is customary to omit the inch (") designation when a scale is specified. The drawing scale may be stated as: FULL SCALE, FULL SIZE, 1:1, HALF SCALE, HALF SIZE, 1/2 SCALE, and so on. The only purpose of the scale is to reproduce the dimensions of a part full size on a drawing or to enlarge or reduce them to some suit-

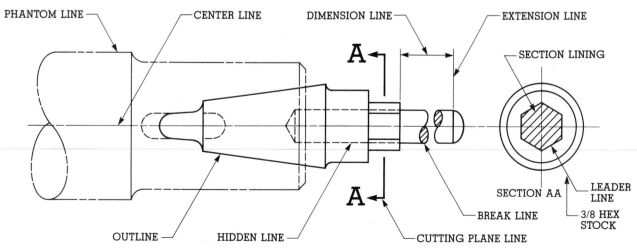

FIGURE 2.24
Examples of line convention.

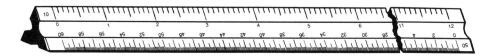

ENGINEER'S SCALE

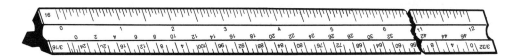

ARCHITECT'S SCALE

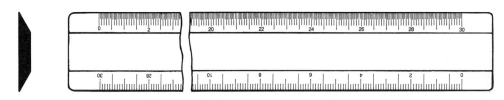

METRIC SCALE

FIGURE 2.25
Triangular and flat scales.

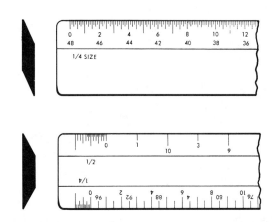

FIGURE 2.26
Mechanical engineer's scale divided into fractions of an inch.

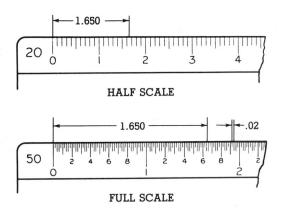

HALF SCALE

FULL SCALE

FIGURE 2.27
Engineer's decimal scale.

TABLE 2.3
Mechanical engineer's fractional scales.

MECHANICAL ENGINEER'S FRACTIONAL SCALES
FULL SIZE (1″ = 1″)
HALF SIZE (½″ = 1″)
QUARTER SIZE (¼″ = 1″)
EIGHTH SIZE (⅛″ = 1″)

TABLE 2.4
Mechanical engineer's decimal scales.

MECHANICAL ENGINEER'S DECIMAL SCALES
FULL SIZE (1.00″ = 1.00″)
HALF SIZE (0.50″ = 1.00″)
THREE-EIGHTHS SIZE (0.375″=1.00″)
QUARTER SIZE (0.25″ = 1.00″)

able size to fit the size of the intended drawing medium.

2.4.1.2 The Metric Scale

Measurement of length in the metric system is based on a unit of distance called a **meter**. A meter is equal to 39.37 inches, or a bit longer than a yard (36 inches). A meter is divided into 100 units identified as **centimeters (cm)**, and each centimeter is divided into ten units identified as **millimeters (mm)**. One hundred centimeters equals one meter and one thousand millimeters equals one meter. For the most part, metric measurements are specified in millimeters for work in technical graphics, just as most mechanical measurements in the English system are specified in inches. Metric scales are graduated and calibrated for their full length, as shown in Figure 2.28, and each scale is identified by a scale ratio such as 1:150 or 1:100 if a drawing scale other than 1:1 is required.

Figure 2.29 illustrates a metric scale with measurements in millimeters (mm). This figure also shows a few sample measurements.

To convert a meter value into millimeters, multiply the given value by 1000, which is the same as moving the decimal point three places to the right. For example:

4.9 meters = how many millimeters?
(4.9)(1000) = 4900 mm

For converting a millimeter value into meters, divide the given value by 1000, which is the same as moving the decimal point three places to the left. For example:

575 millimeters = how many meters?
575/1000 = .575 m

A basic simple reference when working in the metric system of measurement is as follows:

1000 millimeters = 1 meter
100 centimeters = 1 meter
10 millimeters = 1 centimeter

When converting millimeters to inches or inches to millimeters, it is important to remember the following:

25.4 millimeters = 1 inch

FIGURE 2.28
Metric scales.

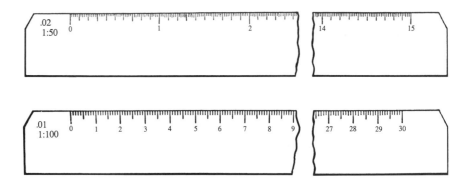

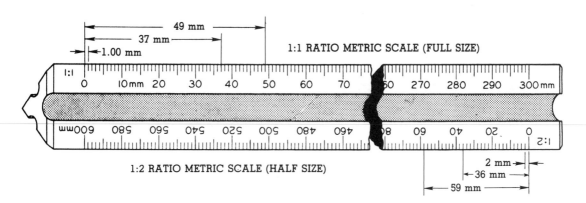

FIGURE 2.29
Millimeter scale and sample measurements.

When required to convert a given millimeter value into inches, divide the millimeter value by 25.4. For example:

295 millimeters = how many inches?
295/25.4 = 11.61 inches

If a given value in inches is to be converted into millimeters, multiply the inch value by 25.4. For example:

4.25 inches = how many millimeters?
(4.25)(25.4) = 107.95 millimeters

When working in fractions, first convert the whole and fractional number to the decimal form, and then multiply the converted value by 25.4 as shown in the example below.

$4\frac{5}{16}$ inches = how many millimeters?
$4\frac{5}{16}$ inches converted to decimal form = 4.3125
(4.3125)(25.4) = 109.54 millimeters

For ease of converting inches to millimeters and millimeters to inches, refer to Appendix A. Using these tables will allow you to convert values easily without performing extensive calculations.

2.4.1.3 The Dual-Dimensioning System

Because the SI system of measurement is not universally accepted, many industries use both the SI and inch-foot systems simultaneously. This combined system is called the **dual-dimensioning system**. When this system is used, both the metric and inch-foot dimensional values are shown. For example, in a method approved by the American National Standards Institute (ANSI), on the field of a drawing, metric dimensions may be shown in brackets [ ] below or to the right of the inch dimensions, as Figure 2.30 illustrates.

Only one of the methods of identification of units in Figure 2.30 is permitted throughout a single drawing. When the method of identifying dual dimensions is selected it must be clearly described in a note near the title block of the drawing, as follows:

Inches

←——————→

[mm]

or

Inches [mm]

The dimension line may be used to separate the inch and metric dimensions. Parentheses ( ) are not to be used as a substitute for brackets [ ] around metric dimensions. Parentheses are reserved for reference dimensions only.

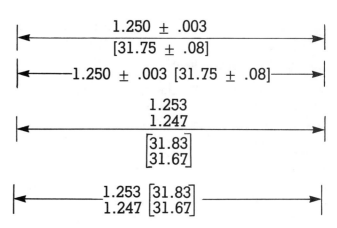

FIGURE 2.30
Acceptable dual-dimensioning practices.

An alternative method of dual dimensioning is the addition on the drawing of a conversion table of all dimensional values, as illustrated in Figure 2.31.

2.4.1.4 The System of International Units (SI)

The International Organization for Standards (ISO) has developed a universally recognized system of well-defined, coordinated units for science and industry which could become the international language of measurement. **SI units** of measurement are divided into three classes in the international system and are authorized by the U.S. Bureau of Standards. They include:

- Base units
- Supplementary units
- Derived units

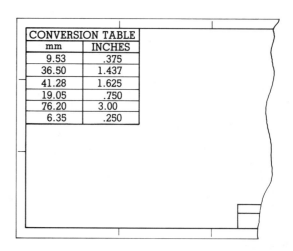

FIGURE 2.31
Alternative method of dual dimensioning.

The seven base units are:

- The meter – the unit of length
- The kilogram – the unit of mass
- The second – the unit of time
- The ampere – the unit of electric current
- The kelvin – the unit of thermodynamic temperature
- The mole – the unit of amount of substance
- The candela – the unit of luminous intensity

The two supplementary units are:

- The radian – the unit of plane angle
- The steradian – the unit of solid angle

These supplementary units are illustrated in Figure 2.32.

The derived units are formed from both base and supplementary units by multiplying, dividing, or raising the unit(s) to a positive or negative power as shown in Appendix A.

There are also several non-SI units that are recognized and that may be used in the SI system. These units are:

- The degree (of angle)
- The degree Celsius (temperature)
- The decibel
- The minute
- The hour
- The day
- The month
- The year
- The liter

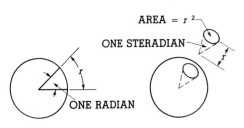

FIGURE 2.32
Radian and steradian defined.

The word **centigrade**, relative to temperature, has been obsolete since 1948. The proper term is **degree Celsius**.

2.5 SUMMARY

The purpose of drafting equipment and tools is to aid the drafter in producing accurate, clear, and legible technical information in a timely and efficient manner.

Both manual and automatic equipment and tools are utilized in the modern drafting laboratory; both were described in detail.

The term **measurement** was defined as the determination or estimation of a value, dimension, or quantity by comparison with some known standard. The major types of measurement systems used throughout the world—the metric, inch-foot (English), dual-dimensioning, and SI systems—were discussed. Practice exercises for this section may be found on pages 2A through 2G in the workbook.

2.6 REVIEW EXERCISES

1. Name two kinds of drafting technology equipment used in the modern drafting laboratory.

2. List four pieces of manual drafting equipment.

3. Name five pieces of automated hardware used in the modern drafting laboratory.

4. Identify two output devices in the automated drafting laboratory.

5. Name seven manual tools used by the drafter.

6. What is the purpose of a template?

7. What is a curve used for?

8. What are three uses for dividers?

9. What kinds of erasers are recommended for use on polyester film?

10. Name three types of media used to produce manual technical graphics.

11. What are the common line widths used on technical drawings?

12. What three types of triangles are used by drafters?

13. What are the standard sizes of drawing paper for "A," "C," "E," "A4," and "A2" drawings?

14. What are the names of the following conventional lines?

a. ——— – ——— – ——— – ——   ———————

b. ————————————————   ———————

c. — — — — — — — — — — — —   ———————

d. ——— — — ——— — — ———   ———————

e. ⌐———→   ┌———→   ———————

15. Is a 3H lead harder than a 6H? Is a 4H lead softer than an H?

16. Name two different types of measurement systems.

17. What are the two basic shapes of scales used for measurement?

18. What is the predominant measurement system used in the United States?

19. What are the primary divisions of a decimal scale?

20. How many inches are there in a meter?

21. How many millimeters (mm) are there in one inch? In 12 inches? In 20 inches?

22. How many inches are there in 254 millimeters? In 560 millimeters? In 784 millimeters?

23. What dimensional values are shown on a drawing when the dual-dimensioning system is used?

24. What are parentheses used for on a drawing?

25. What are the seven base units identified in the SI system?

a.

b.

c.

d.

e.

f.

g.

Section 3

Lettering and Text Presentation

LEARNER OUTCOMES

The student will be able to:

- List three methods for producing text on technical drawings

- Name the style of lettering commonly used on technical drawings

- Explain the six basic steps for producing alphanumeric characters

- Explain the acceptable spacing of words and lines for text material

- Identify two types of freehand and mechanical lettering aids

- Demonstrate learning through the successful completion of review and practice exercises

3.1 TEXT PRESENTATION

The words **printing** and **lettering** have been used for many, many years in drafting technology to indicate alphanumeric characters (letters and numerals) placed on drawings. A more appropriate term used today is **text presentation**. The word **text** has come into prominence with the advent of the computer and more specifically in the field of computer-aided drafting (CAD). Text on technical drawings consists of dimensions, notes, charts, instructions, specifications, legends, lists of materials, and any other information that is best communicated through the use of alphanumeric characters. Text and how it is presented constitute an important part of any drawing.

There are many methods of producing text for technical drawings. Three basic methods used by the drafter will be discussed in this section: the **freehand**, **mechanical**, and **computer-generated** methods.

The American National Standards Institute (ANSI) established its first standard for lettering in 1935, and its most recent specification is entitled Y14.2M-1979, "Line Conventions and Lettering."

3.2 FREEHAND LETTERING TECHNIQUES

Good freehand lettering techniques produce acceptable text material on technical drawings. The ability to produce text that is neat, clear, legible, uniform, and of proper density is a skill that must be acquired by every drafter. All text materials must be of high quality regardless of the final reduction size requirement of the drawing.

Because industry is constantly trying to increase productivity, it is important that a minimum amount of time be spent on lettering. It is important for a drafter to produce text as quickly and as efficiently as possible. Some people have a natural talent for lettering, while others need to work hard to become proficient. Skill in lettering can be acquired only by continued practice. During idle moments, free time, or when riding to or from work, one should, with pencil and pad in hand, practice lettering. Practice is the key.

3.3 STYLES OF LETTERING

Of the several styles of lettering that have been developed over the years, the predominant one for use in drafting technology is the single-stroke, upper-case, commercial Gothic font. **Single stroke** means that

the required thickness or weight of each letter is formed using one stroke; **upper case** indicates that all letters are capitalized; and the **Gothic font** is one in which all strokes of each letter are even or uniform. **Font** means style and is a word that is used frequently today because of its use with computers and computer-aided drafting (CAD) language. A font is a complete assortment of alphanumeric characters of a particular size and style which can be used for the several different applications required by the drafter, artist, or illustrator. Figure 3.1 identifies the more common fonts available.

In this style of lettering or text presentation, all alphanumeric characters are created using the six basic strokes illustrated in Figure 3.2.

Stroke 1 is a single stroke drawn downward from upper left to lower right at an angle of approximately 30 to 60 degrees. **Stroke 2** is drawn downward from upper right to lower left at an angle of approximately 30 to 60 degrees. **Stroke 3** is a vertical line drawn from top to bottom. **Stroke 4** is a horizontal line drawn from left to right. **Stroke 5** is a semicircular stroke that moves downward in a curved, counterclockwise fashion. **Stroke 6** is a semicircular stroke that moves downward in a curved, clockwise fashion (the mirror image of stroke 5). All numerals and all letters of the alphabet can be produced by using combinations of these six strokes.

Either vertical or inclined text is acceptable for use in industry; however, most large companies

A B C D E
a b c d e

A B C D E
a b c d e

A B C D E
a b c d e

A B C D E
a b c d e

A B C D E
a b c d e

A B C D E
a b c d e

FIGURE 3.1
Common fonts.

FIGURE 3.2
Basic strokes for producing single-stroke, upper-case, commercial Gothic font characters.

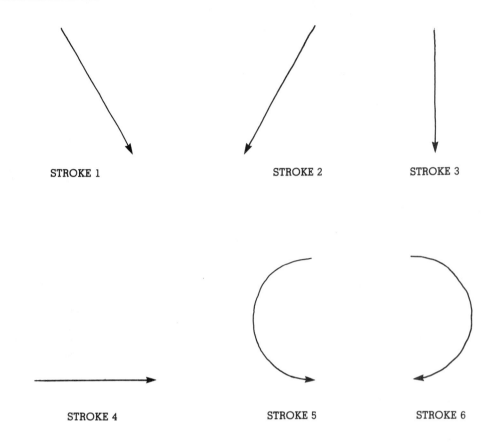

STROKE 1   STROKE 2   STROKE 3

STROKE 4   STROKE 5   STROKE 6

prefer vertical characters. Only one type of lettering, either vertical or inclined, should appear on a given drawing. The method for producing both vertical and inclined letters and numerals is shown in Figure 3.3.

Acceptable and unacceptable lettering practices are illustrated in Figure 3.4.

Fractions and mixed numbers are always drawn using horizontal bars, not angled bars. Acceptable and unacceptable methods of producing fractions and mixed numbers are illustrated in Figure 3.5.

Notes, either general or local, are placed on a drawing so as to align parallel with the bottom of the drawing, as shown in Figure 3.6. This is required for all the practice exercises in the workbook.

3.4 SPACING AND SIZES

Photographic reduction of original drawings or reproduction of drawings on microfilm limits the minimum height of character and minimum line spacing in terms of maintaining legibility. The success of either reproduction process depends on clear, legible, well-formed letters and numerals. Care should be taken to produce open letters and numerals with sufficient space between them to ensure legibility af-

ter reduction. For this reason, vertical spacing between lettered lines should not be less than the height of one letter, and horizontal spacing between words should not be less than one-half the height of one letter (or more than the full height of one letter). Examples of acceptable and unacceptable spacing practices are shown in Figure 3.7.

To provide a basis for text uniformity and consistency, it is recommended that all letter and numeral heights be produced in accordance with those shown in Table 3.1 for practice exercises in the workbook.

3.5 AIDS TO FREEHAND LETTERING

Lettering aids help ensure that freehand lettering on a technical drawing is of good quality by providing uniformity in height and width of characters. When producing freehand lettering, it is necessary to begin with horizontal, vertical, and/or inclined guide lines. An example of a commercially available template for drawing guide lines is the **Ames Lettering Guide**, shown in Figure 3.8(a). This guide is a transparent plastic assembly consisting of two parts: a stationary outer frame and an inner disk that rotates to provide an unlimited number of adjustments for

VERTICAL INCLINED

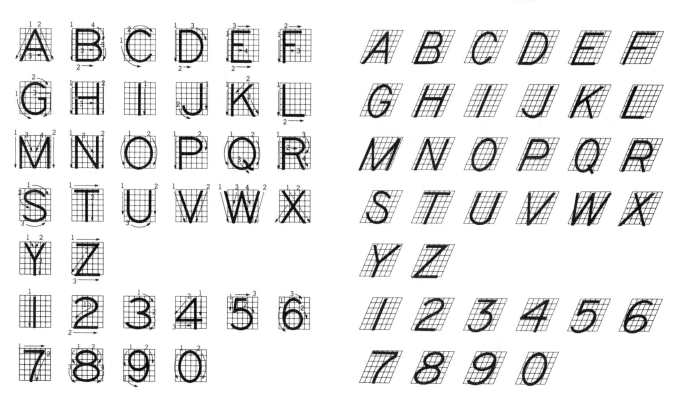

FIGURE 3.3
Method for producing commercial Gothic style lettering.

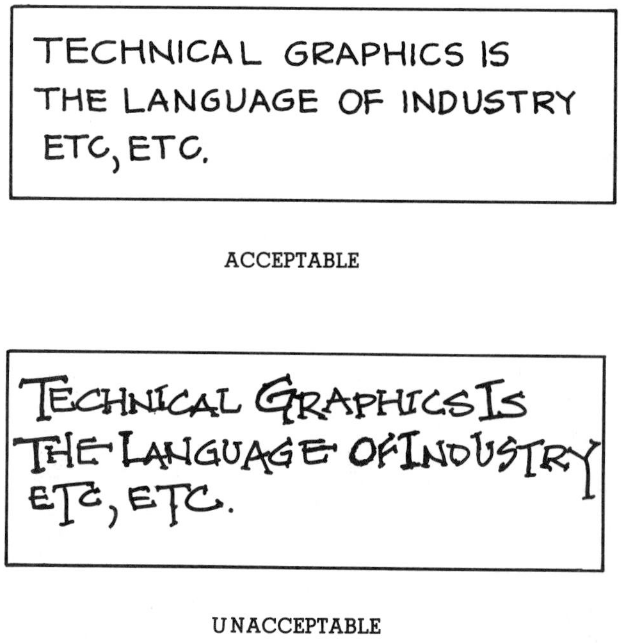

TECHNICAL GRAPHICS IS
THE LANGUAGE OF INDUSTRY
ETC, ETC.

ACCEPTABLE

$2\frac{3}{4}$ $7\frac{5}{8}$ $\frac{13}{16}$ $\frac{15}{32}$

1.562   .3125   1.625

ACCEPTABLE

TECHNICAL GRAPHICS IS
THE LANGUAGE OF INDUSTRY
ETC, ETC.

UNACCEPTABLE

$2\frac{3}{4}$ $7\frac{5}{8}$ $\frac{13}{16}$ $\frac{15}{32}$

1.562 .3125   1.625

UNACCEPTABLE

FIGURE 3.4
Acceptable and unacceptable lettering practices.

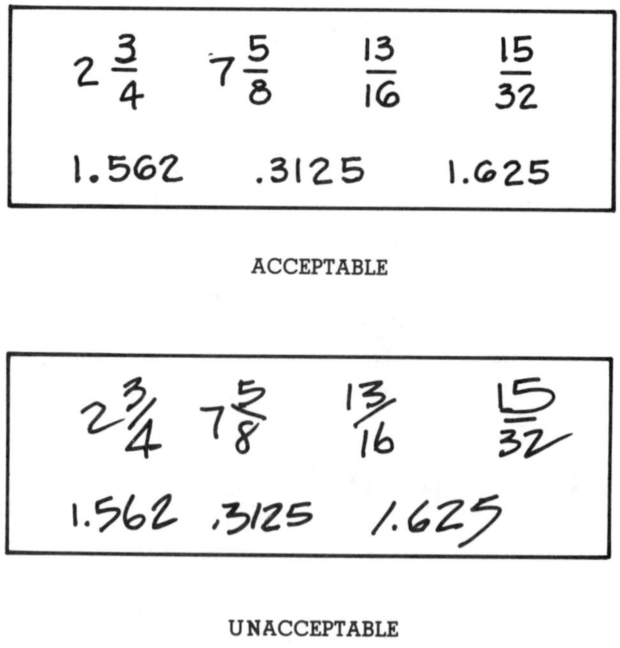

FIGURE 3.5
Acceptable and unacceptable practices for producing
fractions and mixed numbers.

FIGURE 3.6
Alignment of notes.

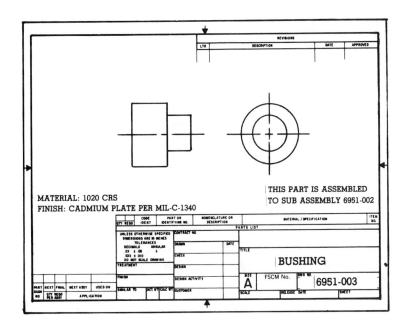

ACCEPTABLE

UNACCEPTABLE

FIGURE 3.7
Acceptable and unacceptable spacing practices.

guide lines. The inner disk contains columns of holes to provide uniformly spaced lines. In addition, the device has the ability to provide direct setting for crosshatching for section lining, and an opening for producing a symbol for surface finishes. Clear, concise instructions for the use of the Ames Lettering Guide are provided in each package. An illustration of the types of guide lines produced by the guide is shown in Figure 3.8(b).

Figure 3.9 shows how to use the Ames Lettering Guide in conjunction with a T-square, a drafting machine, or a straightedge. Guide lines should be very light in weight.

3.6 SUGGESTIONS FOR PRODUCING QUALITY FREEHAND TEXT MATERIAL

There are no specific rules for producing quality text, but there are some basic logical hints that may help. They are:

1. After the technical drawing is complete and before adding any text to the drawing, make a copy of it on the diazo whiteprinter machine, or other available copier.

2. Determine the total amount of text (notes, dimensions, charts, instructions, specifications, etc.) that needs to be added to the drawing. In rough form, add all the required text information to the copy in the approximate locations in which it will appear on the actual drawing.

3. When satisfied as to the completeness and location of the text material, attach the incomplete

TABLE 3.1
Recommended numeral and letter heights.

LOCATION OF LETTERING	NUMERAL AND LETTER HEIGHT - INCHES
DRAWING TITLE, DRAWING NUMBERS, AND OTHER CAPTIONS	.18 HIGH
DRAFTER'S NAME, ALL LETTERING ON FACE OF DRAWINGS, DIMENSIONS, NOTES, TOLERANCES, AND HOLE CHARTS	.12 HIGH

FIGURE 3.8(a)
Ames Lettering Guide.

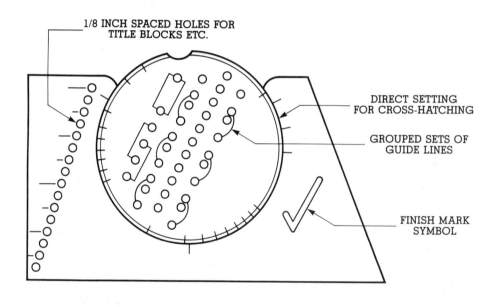

FIGURE 3.8(b)
Vertical, angled, and horizontal guide lines.

original drawing to the drawing surface (with tape, if attached to a drafting board).

4. Using a guide-line template, and a lead holder or pencil with 5H or 6H lead, draw lightly a sufficient number of horizontal and vertical guide lines for all the text material (letters and numerals) required on the drawing. Use Table 3.1 as a reference for determining the heights of characters.

5. After all guide lines for text have been added, remove the tape from the four corners of the drawing. If the drawing is of a convenient size, place the drawing at a comfortable, convenient angle on the drawing surface and begin lettering. If it is a large drawing (22 x 34 inches or larger), it may be more appropriate to leave it attached to the board.

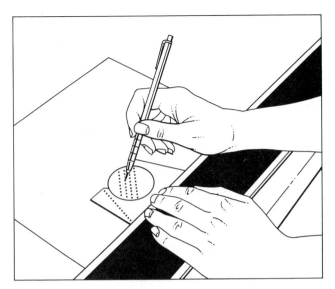

FIGURE 3.9
Use of an Ames Lettering Guide.

FIGURE 3.10
Lettering templates.

6. Re-point (sharpen) lead as often as necessary if a significant amount of text material is to be added.

3.7 MECHANICAL LETTERING AIDS

Several manufacturers produce mechanical lettering aids called **lettering templates**. These templates are used to produce alphanumeric text. They are available in various thicknesses of plastic material and are made in many, many character sizes and fonts to meet the drafter's needs. They produce both vertical and angled lettering. A sampling of available templates for mechanical lettering is shown in Figure 3.10.

Another type of aid is **transfer-type lettering**. Transfer-type lettering is available on translucent sheets with either an adhesive backing or letters that are rubbed off by providing firm and even pressure to the desired area. **Burnishing** (rubbing the letter with a blunt tool) is required to ensure that the transferred characters remain permanently attached. Transfer-type lettering will ensure uniformity and shape of characters, but spacing quality can be determined only by the person performing the transfer operation. Figure 3.11 illustrates examples of available transfer-type lettering.

3.8 COMPUTER-GENERATED TEXT

Any mainframe, mini, or microcomputer that is manufactured today—especially computer-aided drafting (CAD) equipment—has the capability for producing computer-generated text material. Each of these systems is capable of performing word-processing functions. Adding text to technical drawings is a simple task of inputting alphanumeric characters using the keyboard. Text can be inserted and justified to either margin, or can be centered. It also can be placed at any desired angle. Several different fonts and heights can be produced in the **text mode** on the computer. Additional information on computer-generated text material is provided in Section 16.

3.9 SUMMARY

Terms connected with the placement of alphanumeric characters on technical drawings were introduced and defined. The most appropriate term used today is **text**. In drafting technology, text means dimensions, notes, charts, instructions, specifications, legends, and lists of materials added to drawings. Three methods of producing text material on technical drawings were covered: the freehand, mechanical, and computer-generated methods. The predominant style of lettering utilized by industry is the single-stroke, upper-case, commercial Gothic font. Basic strokes for producing Gothic letters were illustrated, and acceptable and unacceptable methods for producing alphanumeric characters were shown. In addition, aids to freehand and mechanical lettering were discussed. Practice exercises for this section are presented on pages 3A through 3D in the workbook.

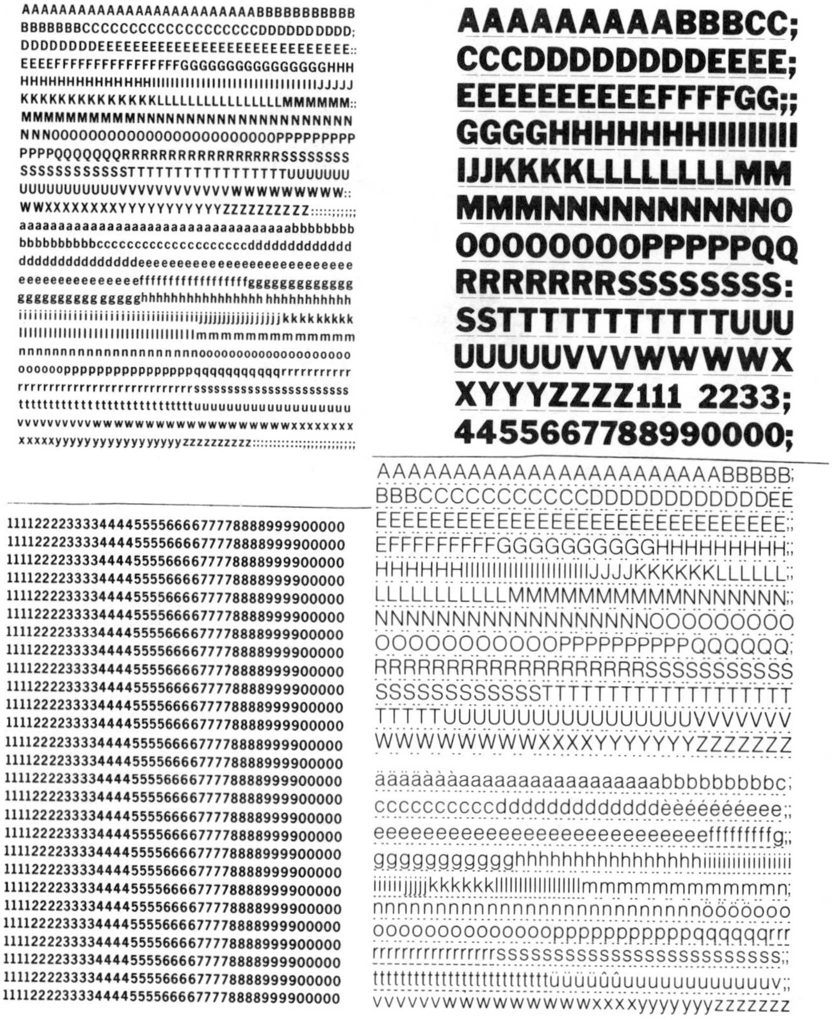

FIGURE 3.11
Transfer-type lettering aids.

3.10 REVIEW EXERCISES

1. Name six different kinds of text material that might be included on a technical drawing.

2. Define alphanumeric characters.

3. Name three basic methods of producing text material.

4. Describe the predominant style of lettering that is used on technical drawings.

5. Explain the six basic strokes used to form Gothic-style alphanumeric characters.

6. Indicate the proper method of showing the following fractions on a technical drawing:

    a. One and five-eighths inches _____

    b. Two and three-sixteenths inches _____

    c. Four and seven-eighths inches _____

    d. Three-fourths inch _____

7. List two types of mechanical lettering aids.

8. Name two uses for templates.

9. Why is the spacing of words and lines important on a drawing?

10. What is the most recent specification for line convention and lettering?

11. Identify five helpful hints for producing freehand text material.

    a.

    b.

    c.

    d.

    e.

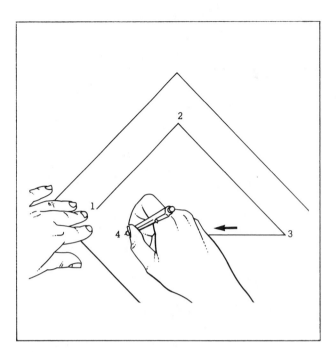

Section **4**

The Technical Sketch

LEARNER OUTCOMES

The student will be able to:

- State the purpose of the technical sketch

- Identify materials required for producing technical sketches

- Identify four types of technical sketches

- Describe the steps required for producing circles and arcs

- List the differences among multiview, isometric, oblique, and perspective sketches

- Demonstrate learning through the successful completion of review and practice exercises

4.1 PURPOSE

The intent of the **technical sketch**, which is always produced freehand, is to communicate a concept, idea, or plan in a rapid but effective manner. Knowing how to interpret and draw technical sketches is an important and desirable skill for the drafter. A technical sketch is often the primary, and sometimes the only, communication message between engineering project personnel and the drafter. Also, it may well be the first drawing the drafter will be required to produce or interpret on a project.

The drafter is not the only person who produces technical sketches. Sketches are drawn by engineers during design and development of new products. Model shop personnel, instrument makers, and engineering model shop builders draw sketches as part of their work. Production and industrial engineers use sketches to show initial or conceptual thoughts on improving production processes. In summary, anyone involved with the engineering and production of goods and services whose communication needs cannot be met by the written or spoken word use and benefit from the technical sketch.

Many preliminary concepts and ideas take the form of a technical sketch that conveys descriptive data and that may later become a finished drawing. Technical sketches, although infrequently drawn to scale, are usually drawn to correct proportions.

The only materials required for producing a technical sketch include a soft grade of lead, a mechanical lead holder or pencil, an eraser, and a drawing medium. The most suitable medium is paper—either a ruled paper with a reproducible or nonreproducible grid of 4, 8, 10, or 12 squares to the inch, or a plain bond (gridless) translucent paper used in conjunction with a grid master underneath it to provide grid lines. An example of a grid format for technical sketching is shown in Figure 4.1.

4.2 TYPES OF TECHNICAL SKETCHES

Several types of technical sketches are used in design and drafting of mechanical equipment. The most frequently used sketches are the **multiview orthographic sketch** and the **isometric sketch**. Others which will be discussed, but which are used to a lesser degree, are the **oblique sketch** and the **perspective sketch**.

4.2.1 The Multiview Sketch

The multiview sketch is also referred to as the **orthographic projection**. In the multiview sketch it is im-

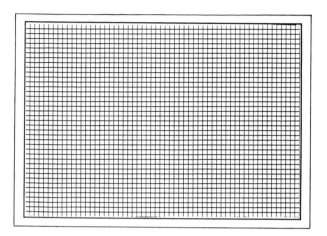

FIGURE 4.1
Grid paper for technical sketching.

portant that views be placed in their proper locations in a method called third-angle projection, where the top view is placed above the front view because both views share the same width dimension. The side view is placed to the right of the front view because both of these views share the same height dimension. The number of views selected depends on the complexity of the object being drawn and the technical opinion of the sketcher. As many as six views, or more, may be required: front, top, right-side, left-side, bottom, rear, and possibly auxiliary views. Figure 4.2 depicts a simple multiview technical sketch on grid paper. Note the locations of the views.

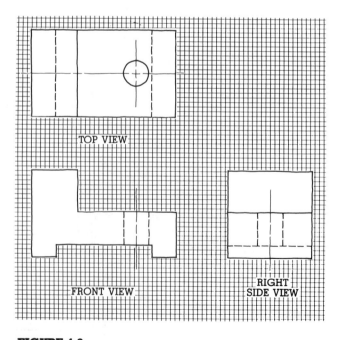

FIGURE 4.2
The multiview sketch.

4.2.2 **The Isometric Sketch**

The **isometric sketch** is a form of **axonometric drawing**. The term "axonometric" implies that an object is presented on a surface so that its perpendicular projections appear as inclined surfaces where three faces are shown. There are two other types of axonometric drawings: the **dimetric drawing**, where two of the three principal faces and axes are equally inclined to the plane of projection; and the **trimetric drawing**, where all three faces and axes of the object make different angles with the plane of projection. Illustrations of axonometric drawings are shown in Figure 4.3.

The isometric sketch is by far the most often used of the three. Isometric lines are drawn vertically and at 30 degrees from the horizontal. The same part that is shown in Figure 4.2 is drawn as an isometric sketch in Figure 4.4. Note that isometric lines are sketched parallel to the 30-degree isometric axis. Vertical and horizontal measurements are scaled directly from the multiview sketch and transferred to the corresponding isometric lines.

4.2.3 **The Oblique Sketch**

In an **oblique sketch**, the receding (depth) lines are usually drawn at 30 degrees, 45 degrees, or 60 degrees to the left or right. This type of sketch involves a combination of a flat front surface with the depth lines receding at the selected angle. The front surface projects the true size and shape of the object. Figure 4.5 illustrates the same object as that shown in Figures 4.2 and 4.4, but drawn as an oblique sketch.

4.2.4 **The Perspective Sketch**

The **perspective sketch** is a form of projection in which the projecting or extended lines radiate from a point of sight called the vanishing point. It shows the relative proportions of the object as the human eye views them, and it provides the most realistic view of an object of all the various types of projections. Figure 4.6 shows a perspective sketch of the same object shown in Figures 4.2, 4.4, and 4.5.

4.3 **SKETCHING SQUARES, RECTANGLES, AND TRIANGLES**

Sketching squares, rectangles, and triangles is as simple as drawing horizontal lines. For most people, sketching horizontal lines is easier than drawing vertical or angled ones. Therefore, try to make certain that most, if not all, of the lines of the square, rectangle, or triangle are drawn horizontally, as illustrated

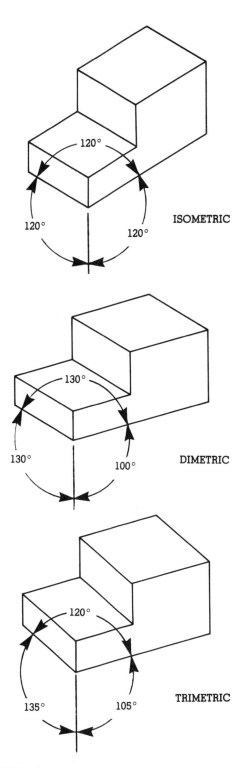

FIGURE 4.3
Axonometric drawings.

in Figure 4.7(a), (b), and (c). To accomplish this task, it will be necessary to rotate the medium several times until the sketch is complete. Horizontal lines may be drawn from left to right or from right to left, whichever is more comfortable.

FIGURE 4.4
The isometric sketch.

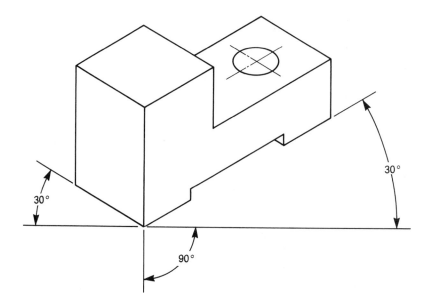

FIGURE 4.5
The oblique sketch.

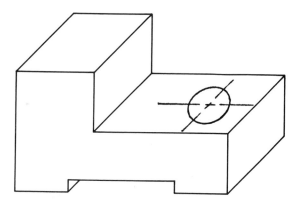

4.4 SKETCHING CIRCLES AND ARCS

Although sketching circles and arcs can be particularly difficult, the methods illustrated in Figure 4.8 will assist in the process. These methods require that

both vertical and horizontal lines be drawn first, crossing at a point where the center of the circle will be located. Next, try to visualize the diameter of the circle that needs to be drawn. Translate this mental picture into equally spaced radial points or short dash lines. Try to plot a minimum of four or as many as 12 equally spaced points. Using several points will ensure a smoother, more perfect circle. Finally, connect the radial points with an even, constant pressure of the lead on the drawing medium. Completing an arc equivalent to one quadrant (one-fourth) of the circle and then rotating the drawing until all four quadrants are complete may prove beneficial, as the figure shows.

4.5 SUGGESTIONS FOR PRODUCING THE TECHNICAL SKETCH

The quality of a technical sketch depends on several factors such as drafting style, experience of the

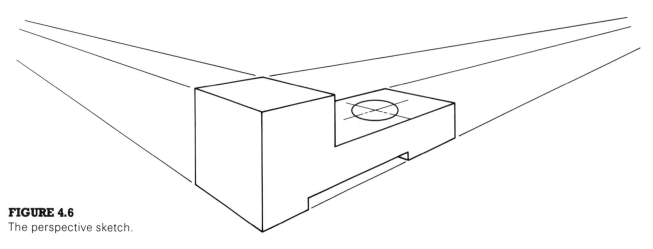

FIGURE 4.6
The perspective sketch.

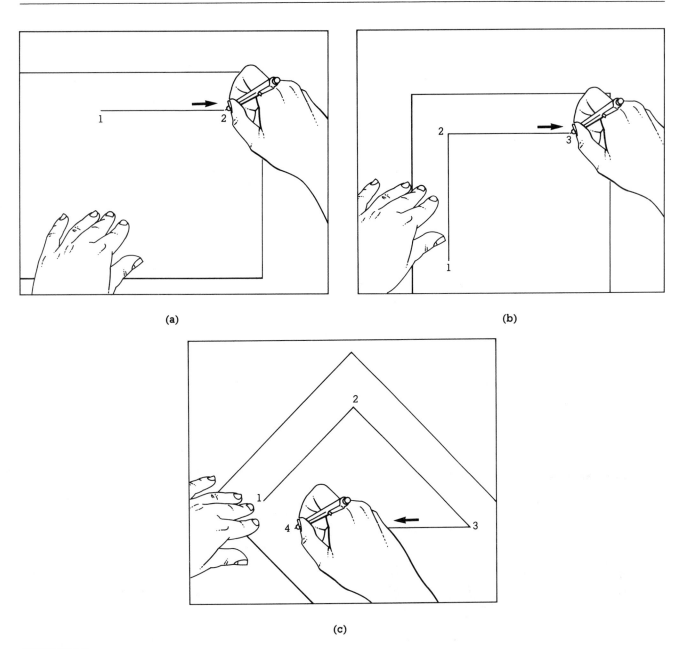

(a)

(b)

(c)

FIGURE 4.7
Producing a technical sketch.

drafter, steadiness of hand, and use of proper materials. The following suggestions are offered for producing quality technical sketches.

1. Select the proper drawing medium and lead for sketching based on the size, complexity, and reproducibility requirements of the drawing.
2. Determine the type of sketch required. Will it be a multiview, isometric, oblique, or perspective drawing? Will it require dimensions and/or text material?
3. Do not attach the drawing medium to the drawing surface, so that the medium can be placed at any

comfortable angle for sketching. Lines can be drawn straighter, more smoothly, and more accurately if as much of the arm as possible is supported by the drafting board surface. Do not allow the unsupported arm to hang off the edge of the drawing board or table when sketching.

4. Grip the pencil or lead holder firmly and maintain a comfortable angle between the drawing surface and the pencil or lead holder.
5. Regardless of the type of sketch required, try to produce it using as many horizontal lines as possible, rotating the drawing medium as often as necessary until completion.

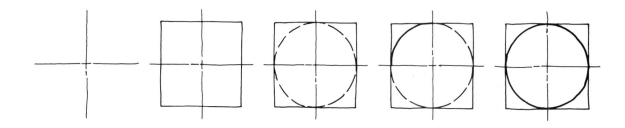

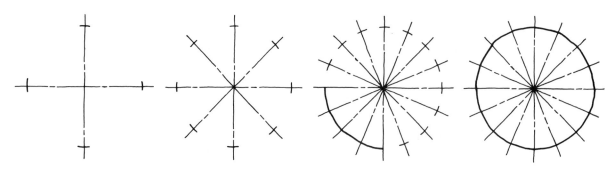

FIGURE 4.8
Sketching circles and arcs.

4.6 SUMMARY

The technical sketches most frequently utilized in industry are the multiview sketch, the isometric sketch, the oblique sketch, and the perspective sketch. The technical sketch provides an easy method for communicating an idea, concept, or plan in a rapid but effective manner. Technical sketches are produced by anyone whose needs for communi-

cation cannot be met by the written or spoken word. Methods for producing various types of technical sketches were introduced. In addition, details on how to sketch squares, rectangles, triangles, circles, and arcs were covered. Finally, a step-by-step process for producing a quality technical sketch was outlined. Practice exercises for this section may be found on pages 4A through 4H in the workbook.

4.7 **REVIEW EXERCISES**

1. What is the purpose of the technical sketch?

2. What basic materials are required for producing technical sketches?

3. What four types of technical sketches are used in industry?

4. Describe the three basic views in the multiview sketch and their relative locations.

5. What are the three types of axonometric sketches?

6. At what angle are isometric lines drawn?

7. Sketch proportionately correct isometric drawings of the following:

a.

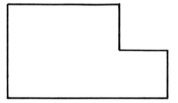

b.

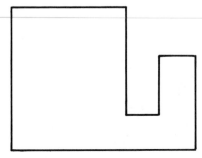

8. At what angle are lines drawn in the oblique sketch?

9. Produce proportionately correct oblique sketches of the following:

a.

b.

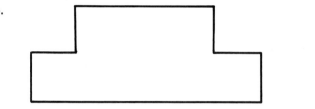

10. Describe the perspective sketch.

11. Identify the steps in producing circles and arcs.

12. List the five suggestions for producing a quality technical sketch.

    a.

    b.

    c.

    d.

    e.

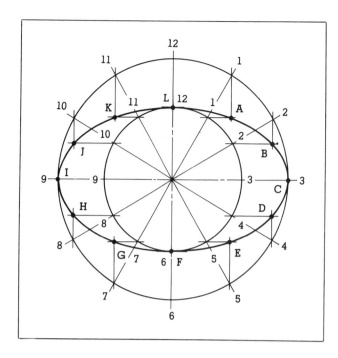

Section 5

The Geometric Construction

LEARNER OUTCOMES

The student will be able to:

- Explain the need to produce geometric constructions

- Define several terms used in the production of geometric constructions

- Identify several shapes and forms of triangles, regular polygons, and quadrilaterals

- List the steps in producing solutions to various geometric problems

- Identify suggestions recommended for producing geometric constructions

- Demonstrate learning through the successful completion of review and practice exercises

5.1 PURPOSE

The **geometric construction** forms the basis for all technical drawings. The purpose of the geometric construction is to accurately develop plane geometric shapes ranging from squares, triangles, and three-dimensional cylinders to complex irregular curves and ellipses. These constructions are normally produced without the aid of a scale, but rather with simple drafting tools.

Engineers, designers, and drafters regularly perform the task of producing geometric constructions in their work, applying the principles of plane geometry. The process involved in the production of geometric constructions requires a basic understanding of plane geometry. Geometric construction skills can be acquired and demand precision and the correct use of drafting instruments. In developing the various geometric constructions in this section it is important that drafting tools be in good repair. 4H to 6H leads are normally used for constructions that produce very light-weight lines. These lines need not necessarily be erased when the construction is completed. A small error or inaccuracy in the solving of a geometric problem could result in a serious error in the final construction.

5.2 DEFINITIONS OF TERMS

Knowledge and use of the following terms are important to the solution of geometric construction problems in this section.

- **Line**   A mark drawn with a pencil or pen, in any direction. A straight line is the shortest distance between two points. Lines may be drawn thin or thick depending on the application, as shown in Figure 5.1(a).
- **Point**   A precise position in space whose location is normally shown as an intersection of two very short lines. A dot may be used as a temporary point location. A **crosshair** is preferred, as illustrated in Figure 5.1(b).
- **Angle**   The space that is formed within two lines originating at a common point. There are three types of angles: the **acute angle**, whose space occupies less than 90 degrees; the **right angle**, whose space occupies exactly 90 degrees; and the **obtuse angle**, whose space occupies more than 90 degrees. Illustrations of each are shown in Figure 5.1(c).
- **Circle**   A closed plane curve in which all points are at an equal distance from the center. A circle

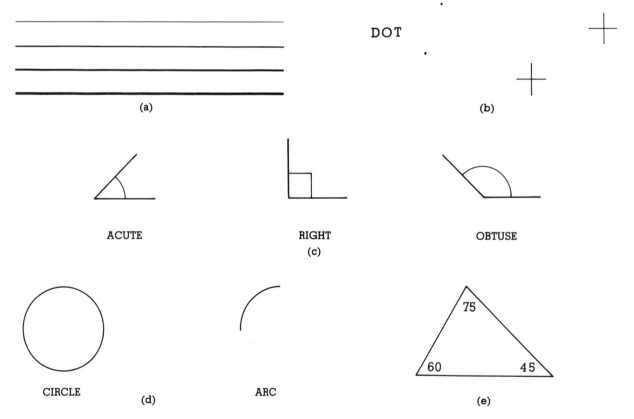

(a)

DOT
(b)

ACUTE   RIGHT   OBTUSE
(c)

CIRCLE   (d)   ARC

75
60   45
(e)

FIGURE 5.1
(a) Lines; (b) Points; (c) Angles; (d) Circle and arc; (e) Triangle.

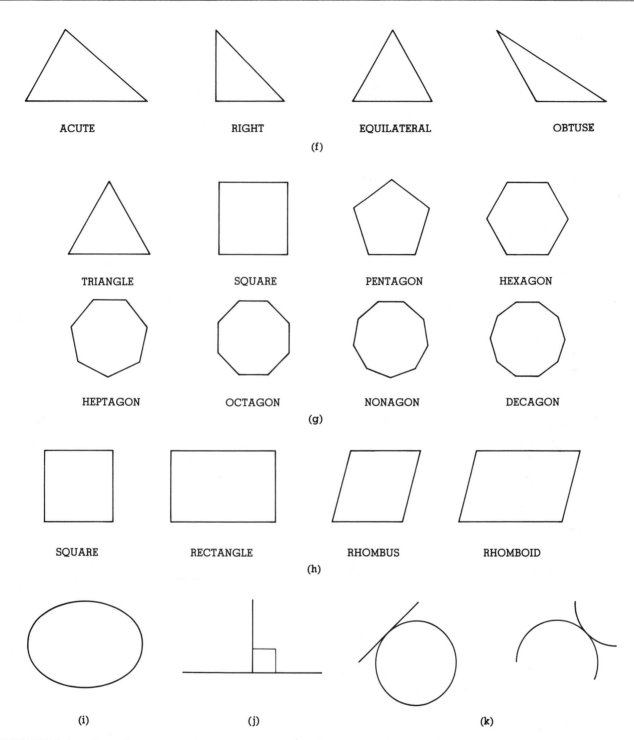

FIGURE 5.1 (continued)
(f) Commonly used triangles; (g) Common regular polygons; (h) Common types of
quadrilaterals; (i) Ellipse; (j) Perpendicular lines; (k) Tangent lines.

contains 360 degrees. An **arc** is a segment of a
circle. Each is illustrated in Figure 5.1(d).
■ **Triangle** A closed plane figure having three sides
and three interior angles. A triangle's three angles
always total 180 degrees, as depicted in Figure
5.1(e). Commonly used triangles are shown in
Figure 5.1(f).

■ **Polygon** A closed plane figure having three or
more straight sides. A **regular polygon** is a figure
whose sides are all equal in length. Common polygons include the triangle, the square, the pentagon, the hexagon, the septagon, the octagon, the
nonagon, and the decagon. Figure 5.1(g) represents the eight commonly used regular polygons.

- **Quadrilateral** A polygon with four sides. Four types whose opposite sides are parallel, which identifies each as a **parallelogram**, are depicted in Figure 5.1(h).
- **Ellipse** A closed symmetrical curve that resembles a flattened circle. It is formed by a point moving so that at any position the sum of the distances from two fixed points (foci) is a constant (equal to the major diameter). The terms **minor diameter**, **major diameter**, **minor radius**, and **major radius** are used when producing an ellipse. Very often ellipses are required to represent circles, holes, and round areas on oblique, isometric, and inclined surfaces. It is the drafter's responsibility to see that the ellipse is constructed correctly. Figure 5.1(i) shows a typical ellipse.
- **Perpendicular** The meeting of a given line or surface at exactly right angles (90 degrees) with another line or surface. Figure 5.1(j) illustrates two lines meeting at 90 degrees.
- **Tangent** A line or curve that touches (without crossing) at a single point of a circle or an arc. Examples are represented in Figure 5.1(k).

There are a number of basic geometric constructions with which the drafter should be familiar. The beginning drafter should follow the basic sequence in each illustration that follows to learn how to develop the various geometric forms. In each of the examples there is **given** information that is basic to the solution.

5.3 BISECTING A LINE

To **bisect** a line means to find its center point or to divide it in half. The stated problem is to bisect a line by the compass method utilizing the step-by-step procedure outlined below and illustrated in Figure 5.2.

Given: Line AB in Figure 5.2.
Step 1: Using points A and B as center points for radii, set radius on compass greater than one-half of AB. Strike radius AC and radius BD.
Step 2: Draw a straight line through points C and D. Line CD bisects line AB and is called the **perpendicular bisector** because it forms an exact 90 degree angle with line AB.

5.4 BISECTING AN ACUTE ANGLE

Given: Acute angle ABC in Figure 5.3.
Step 1: Set compass point at B, and produce radius BD so that it crosses BC and BA at points E and F.

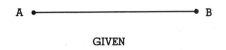
GIVEN

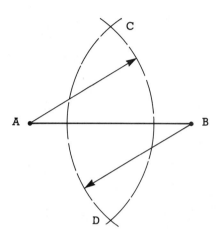
STEP 1

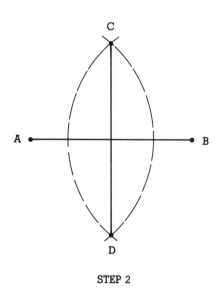
STEP 2

FIGURE 5.2
Procedure for bisecting a line.

Step 2: With the compass, strike equal arcs from points E and F whose radius r is greater than one-half of the distance AC. Make certain that the arcs intersect at G.
Step 3: Draw a straight line through B and G. BG bisects angle ABC, producing two equal angles, ABG and CBG.

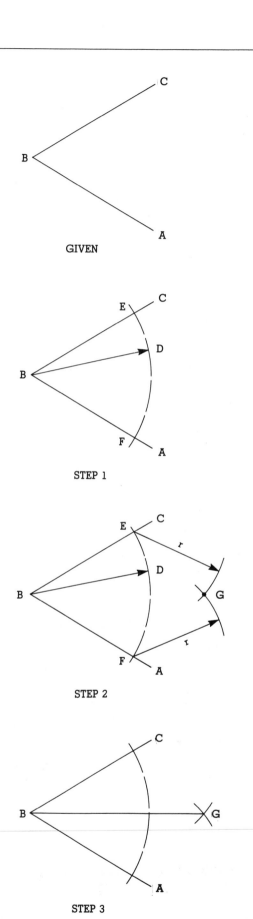

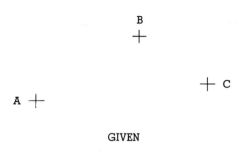

GIVEN

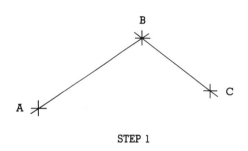

STEP 1

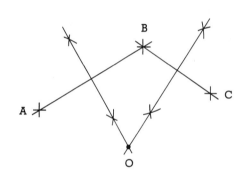

STEP 2

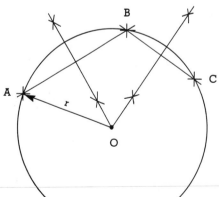

STEP 3

FIGURE 5.3

Procedure for bisecting an acute angle.

FIGURE 5.4

Procedure for constructing a circle through three given points.

5.5 CONSTRUCTING A CIRCLE THROUGH THREE GIVEN POINTS

Given: Three randomly located points A, B, and C in Figure 5.4.

Step 1: Draw a straight line through points A and B and another straight line through points B and C.

Step 2: Using the method for bisecting a line in Figure 5.2, bisect lines AB and BC. Extend bisectors until they intersect at point O. Point O is the center of the intended circle.

Step 3: Set compass point at O and strike radius OA, OB, or OC, which should all be equal in length. Complete radius until a full circle is formed.

5.6 CONSTRUCTING A LINE PARALLEL WITH A GIVEN LINE

Given: Line AB in Figure 5.5.

Step 1: Place compass point on line AB, closer to point A than to point B. Strike any convenient radius r.

Step 2: Using the same radius r, strike a second radius from a point closer to point B on line AB.

Step 3: Draw a straight line through the tangents (high points) of both radii r. The resulting line CD is parallel to line AB.

5.7 CONSTRUCTING A PERPENDICULAR TO A LINE FROM A POINT NOT ON THE LINE

Given: Line AB and random point P in Figure 5.6.

Step 1: Using point P as the center, strike a convenient radius r to intersect line AB at points C and D.

Step 2: Strike two equal arcs significantly greater than r, from points C and D so that they intersect each other at point E.

Step 3: Draw a straight line from point E through point P and extend it until it touches line AB. This line is perpendicular to line AB, which means that it forms a 90-degree angle with line AB.

5.8 DIVIDING A STRAIGHT LINE INTO EQUAL PARTS

Given: Straight line AB in Figure 5.7, to be divided into five equal parts.

Step 1: Draw line BC perpendicular to AB.

Step 2: Place a 0 at point A. Place a scale at the 0 point and pivot the scale until the five-inch point on the scale crosses line BC. Place short vertical

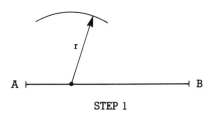
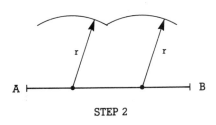
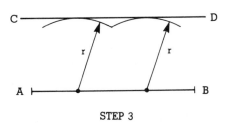

FIGURE 5.5
Procedure for constructing a line parallel with a given line.

dashes at the one-, two-, three-, and four-inch marks on the scale on the five-inch-long line. Identify these inch marks as 1, 2, 3, and 4, and place a 5 at point C.

Step 3: Project lines downward from the points marked 1, 2, 3, and 4 to cross line AB, making certain they are parallel to BC. The original line AB has now been divided into five equal parts. This method can be used to divide a line of any length into the desired number of equal parts.

5.9 CONSTRUCTING AN ANGLE FROM A GIVEN ANGLE

Given: Angle ABC formed by two straight lines that meet at B, and radius r, which intersects BA at G and BC at F. These are shown in Figure 5.8.

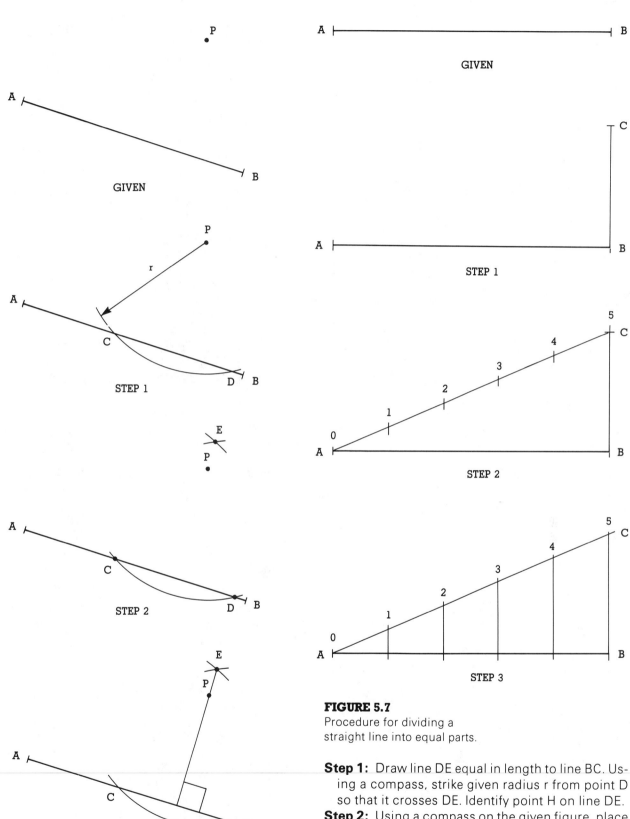

FIGURE 5.6
Procedure for constructing a perpendicular to a line from a point not on the line.

FIGURE 5.7
Procedure for dividing a straight line into equal parts.

Step 1: Draw line DE equal in length to line BC. Using a compass, strike given radius r from point D so that it crosses DE. Identify point H on line DE.

Step 2: Using a compass on the given figure, place compass point at F and strike an arc at G. Transfer arc FG to the figure where the angle is to be transferred. It will become arc HJ.

Step 3: Draw a straight line through DJ. Angle JDH is equal to angle ABC.

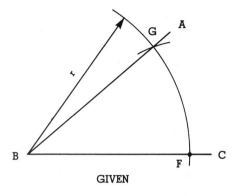

GIVEN

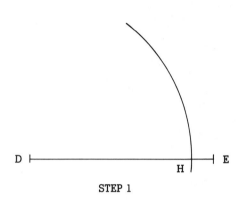

STEP 1

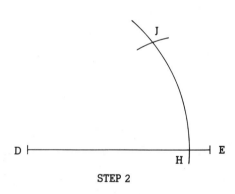

STEP 2

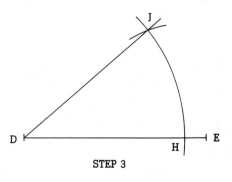

STEP 3

FIGURE 5.8
Procedure for constructing an angle from a given angle.

5.10 CONSTRUCTING A PENTAGON

Given: A circle with horizontal and vertical center-lines AB and CD that will circumscribe (go around) a pentagon (see Figure 5.9). The center of the circle is O.

Step 1: Draw the perpendicular bisector of line OB. Using the midpoint E as the center, strike radius r to C to form arc CF.

Step 2: Place compass point at C and draw a radius through F to produce arc FG.

Step 3: From point C, strike radius CG, which also goes through point F. Using this radius and starting at G, strike arcs GH, HI, IJ, and JC. Draw straight lines from C to G, G to H, H to I, I to J, and J to C. If all the construction work was performed accurately, this should produce the required pentagon.

5.11 CONSTRUCTING A REGULAR HEXAGON, GIVEN THE DISTANCE ACROSS FLATS

Given: The distance across flats, AB, in Figure 5.10.

Step 1: Draw a horizontal center line through AB and a perpendicular bisector center line of AB called CD, whose center is O.

Step 2: Placing the compass point at O, strike radius OA until it makes a complete circle. This should be a very light-weight line.

Step 3: With a 30/60-degree triangle and a T-square, or by using a drafting machine, draw tangent lines as shown that establish the six sides of the hexagon (with opposite sides parallel and all sides equal in length).

This construction practice is used in industry for drawing special hexagonal head bolts, screws, and nuts when templates are not available.

5.12 LOCATING THE CENTER OF A GIVEN CIRCLE

Given: A circle without a center point (see Figure 5.11).

Step 1: Using a T-square, straightedge, or drafting machine, lay out a horizontal line approximately halfway from the estimated center to the top of the given circle. Identify points A and B where the horizontal line intersects the circle.

Step 2: Drop perpendicular lines from points A and B. Identify points C and D where these perpendicular lines intersect the circle.

Step 3: Draw straight lines from point A through point D and from point C through point B. The

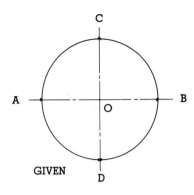

GIVEN

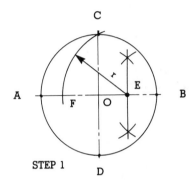

STEP 1

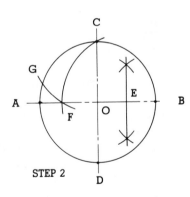

STEP 2

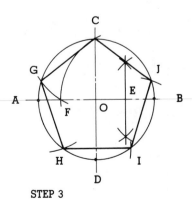

STEP 3

FIGURE 5.9
Procedure for constructing a pentagon.

GIVEN

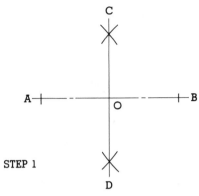

STEP 1

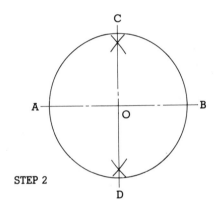

STEP 2

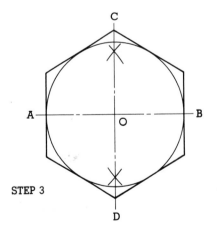

STEP 3

FIGURE 5.10
Procedure for constructing a regular hexagon, given the distance across flats.

point of intersection of these two lines, point O, represents the true center of the given circle. To check the accuracy of the problem solution, place compass point at O and swing radius OA for 360 degrees to make certain that OA equals OB, OC, and OD.

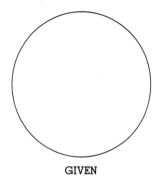

GIVEN

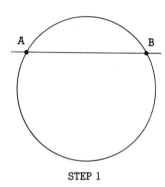

STEP 1

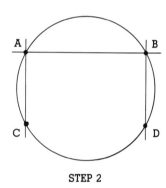

STEP 2

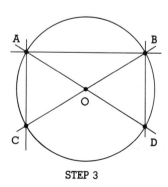

STEP 3

FIGURE 5.11
Procedure for locating the center of a given circle.

5.13 LOCATING TANGENT POINTS

The method for determining tangent points is essentially the same for the right angle, the acute angle, and the obtuse angle. The procedure outlined below, and illustrated in Figure 5.12, is for constructing an arc tangent to a right angle.

Given: Two lines AB and BC, which are at right angles to each other, and radius r (see Figure 5.12).
Step 1: Extend the compass to the length of the given radius, r. Strike this radius twice from line AB and twice from line BC, as shown.
Step 2: Draw a straight line across the tangent (high) points of the arcs from line AB. Do the same with the arcs from line BC.
Step 3: The point of intersection O of the two lines drawn across the tangent points of the arcs is the center for the radius that will be tangent to the right angle formed by lines AB and BC. ·

5.14 CONSTRUCTING AN OGEE CURVE

An **ogee curve**, which is also referred to as a reverse S-curve, is used to connect two parallel lines. This construction practice is used in determining bends in pipes and curvatures in rail and road construction.

Given: Two parallel lines AB and CD offset from each other, as shown in Figure 5.13.
Step 1: Draw a straight line through points B and C. Construct the perpendicular bisector of line BC to establish point O.
Step 2: Construct perpendicular bisectors of lines BO and CO.
Step 3: Construct a perpendicular at point B to intersect with the perpendicular bisector of line BO. This intercept is point E. Construct a perpendicular at point C to intersect with the perpendicular bisector of line CO. This intercept is point F.
Step 4: With the compass point at E, strike radius EB. Radius EB should be tangent to line AB and should cross line BC at point O. Radius EB is the same as radius FC. Set compass point at F and strike radius FC. This completes the ogee curve.

5.15 CONSTRUCTING AN ELLIPSE

Although a template is most often used, there are several methods for constructing an ellipse, including the true method and the approximate method (the true method is slightly more accurate than the approximate method). The designer and the drafter need to know how to construct an ellipse to truly pro-

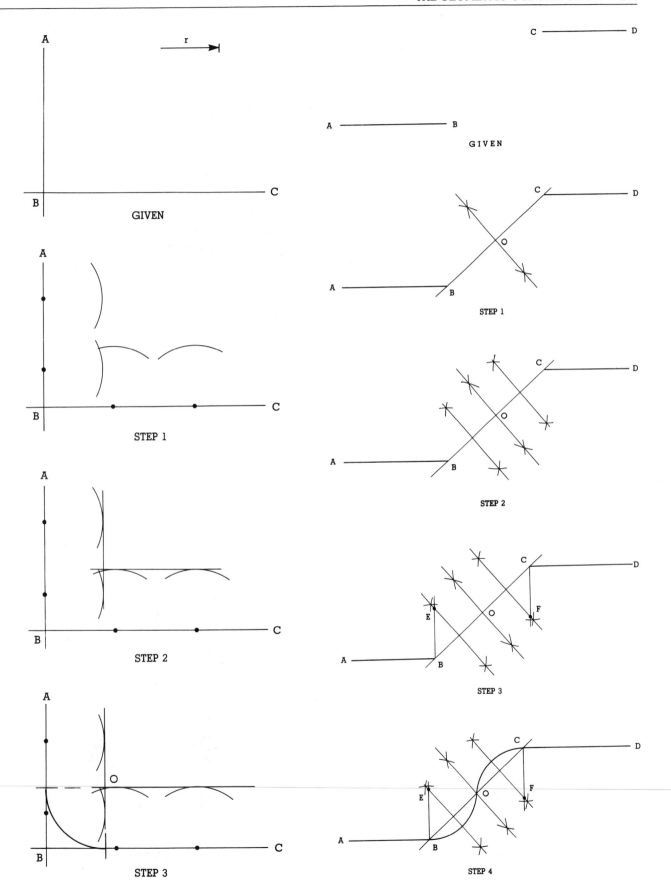

FIGURE 5.12
Procedure for locating tangent points.

FIGURE 5.13
Procedure for constructing an ogee curve.

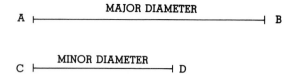

MAJOR DIAMETER

MINOR DIAMETER

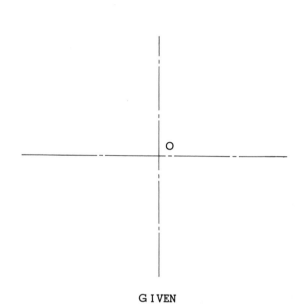

GIVEN

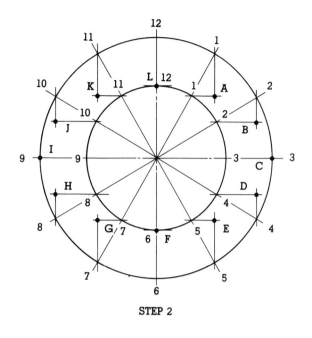

STEP 2

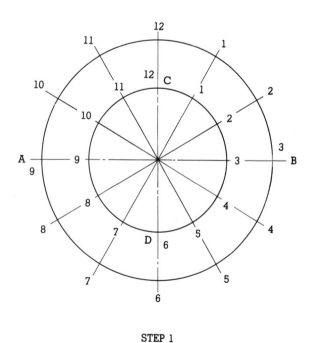

STEP 1

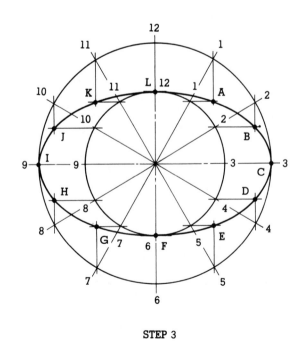

STEP 3

FIGURE 5.14
Procedure for constructing an ellipse.

ject holes, slots, curves, and radii on inclined surfaces. Two important terms used in ellipse construction are **major diameter** and **minor diameter**. The procedure described below is referred to as the **two-circle method** of constructing an ellipse and is illustrated in Figure 5.14.

Given: Major diameter AB, minor diameter CD, and horizontal and vertical center lines intersecting at point O (see Figure 5.14).

Step 1: Lay out two concentric circles (using the same center point, O) with diameters equal to the given major diameter AB and minor diameter CD. Draw light lines that will divide the circles into 12 equal parts by using a 30/60-degree triangle or a drafting machine. Number the intersected points 1 through 12 on the major and minor diameters.

Step 2: Using a triangle and a T-square, or a drafting machine, draw vertical lines downward from points 1, 2, 10, and 11 and upward from points 4, 5, 7, and 8 on the major diameter. Draw horizontal lines from the points where the 12 lines intersect the minor diameter. Extend these horizontal lines until they intersect the vertical lines previously drawn from the numbered points on the major diameter. Label these points A through L, as shown in Figure 5.14, Step 2.

Step 3: Using an irregular curve, draw very lightly the ellipse defined by points A through L. When satisfied that the formed ellipse is as smooth as possible, darken the line. This line is an ellipse constructed using the two-circle method.

5.16 SUGGESTIONS FOR SOLVING PROBLEMS INVOLVING THE GEOMETRIC CONSTRUCTION

The goals for solving any geometric construction problem are to produce the required construction using simple geometric principles, to maintain accuracy and neatness in the construction, and to produce it in the minimum amount of time.

The steps identified below should be followed when solving geometric construction problems.

1. Select the drawing medium on which the construction will be produced.
2. Review the problem to be solved. Be sure to identify the **given** information. The starting point for solving any geometric construction problem is to determine what is given.
3. Using a sharp lead in the lead holder and/or the compass, follow the step-by-step procedure in the text, making certain that all construction lines are drawn very lightly.
4. Do all that is required at each step before proceeding to the next. Identify important point locations on the construction after each step.
5. After completing the last step, examine the solution to make certain it is constructed properly and accurately.
6. When satisfied with the solution to the problem, darken all lines that represent the solution, using whatever lead and line weight are most appropriate. Do not erase construction lines.

5.17 SUMMARY

Geometric construction skills are important to design and drafting personnel. These skills can be obtained through the application of the basic principles of plane geometry and require precise and appropriate use of technical drafting instruments. Several terms that are useful for solving problems in geometric construction were defined. Different examples of commonly used geometric constructions were presented, and a step-by-step process for each was explained and detailed. Finally, suggestions for solving problems involving geometric constructions were outlined. Practice exercises for this section are presented on pages 5A through 5H in the workbook.

5.18 REVIEW EXERCISES

1. What is the purpose for producing the geometric construction?

2. Define the following terms:

Line:

Point:

Polygon:

Ellipse:

3. What is an acute angle?

4. How many degrees are there in a circle?

5. How many degrees are there in a triangle?

6. How many sides does a pentagon have? An octagon? A decagon?

7. Name three types of quadrilaterals.

8. What does "to bisect" mean?

9. Give two practical examples of where the ogee curve is used.

10. Why should the designer and drafter know how to construct an ellipse?

11. What goals should be kept in mind for solving geometric construction problems?

12. List six recommended steps for successfully solving geometric construction problems.

   a.

   b.

   c.

   d.

   e.

   f.

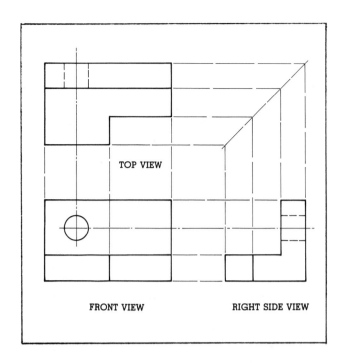

TOP VIEW

FRONT VIEW   RIGHT SIDE VIEW

Section 6

The Multiview Drawing

LEARNER OUTCOMES

The student will be able to:

- Identify the purpose and function of the multiview drawing

- Describe the differences between third-angle and first-angle projection

- State the importance of the selection and placement of views

- List six view placement principles

- Describe the differences between lines and surfaces

- Explain the precedence of lines concept

- Identify the recommended steps in producing a multiview drawing

- Demonstrate learning through the successful completion of review and practice exercises

6.1 PURPOSE

Of all the different types of drawings produced in engineering departments around the world, none is used more extensively than the **multiview drawing**. It is the major type of technical drawing used by industry. The purpose of the multiview drawing is to represent the various faces of an object in two or more views on planes at right angles to each other by extending perpendicular projection lines from the object to these planes. The multiview drawing allows the observer to view an object from more than one orientation. The **top view**, for example, illustrates how the object appears when viewed from above; the **front view** shows how the object appears when viewed from the front; and the **side view** represents the object's features when viewed from the side. Depending on the complexity of the object, it may require one view or as many as six or more. Each view is in direct relationship with the next, but each view portrays the object from a different orientation. Figure 6.1 provides an example of a multiview drawing.

6.2 THE THIRD-ANGLE ORTHOGRAPHIC PROJECTION

The predominant method for producing multiview drawings in the United States and Canada is called **third-angle orthographic projection**. In third-angle orthographic projection, different views of an object (front, top, and side) are systematically arranged on

a drawing medium to convey necessary information to the reader. Features are projected from one view to another. The term "orthographic" has its origin in Greek history, from the words **orthos**, which means straight, correct, at right angles to, and **graphikos**, which means to write, or to describe by drawing lines.

To better understand the theory behind this type of orthographic projection, visualize the transparent box in Figure 6.2, with the object to be drawn placed inside the box so that its axes are parallel to the axes of the box. The projections on the sides of the box are the views one sees by looking straight at the object through each side. If each view is drawn as seen on the side of the box and the box is unfolded and laid flat, as illustrated in Figure 6.3(a), the result is a six-view orthographic projection, better known as a multiview drawing. Any of the six views could be regarded as the **principal view**, and the box could be folded out from one view as easily as from another. When the box is unfolded from the front view of the object, as shown in Figure 6.3(b), it is called **third-angle projection**.

6.3 THE FIRST-ANGLE ORTHOGRAPHIC PROJECTION

First-angle orthographic projection is the preferred method for drafting in most European and Eastern countries. The major difference between the third-angle and first-angle methods is how the object is projected and the positions of views on the drawing. In the third-angle drawing, the projection plane is observed to be between the viewer and the object with the views being projected forward to that plane.

In first-angle projection, the projection plane is observed to be on the far side of the object from the viewer. Figure 6.4 illustrates the first-angle projection concept. Note how the views are projected to the rear and onto the projection plane instead of being projected forward. Compare Figure 6.3(a) and (b) with Figure 6.4 to visualize the differences between first- and third-angle orthographic projection.

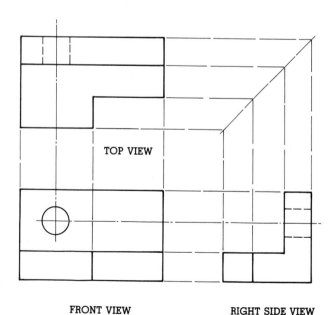

TOP VIEW

FRONT VIEW   RIGHT SIDE VIEW

FIGURE 6.1
Multiview drawing.

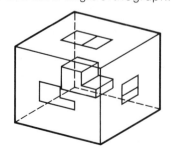

FIGURE 6.2
Transparent box encompassing object to be drawn.

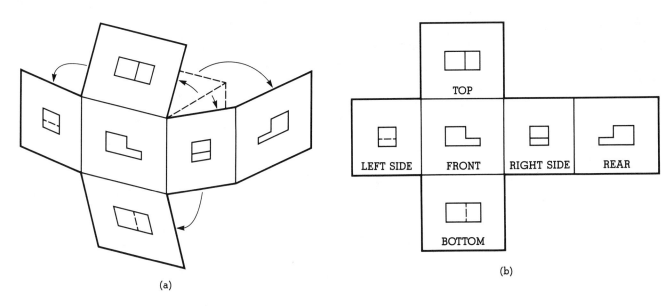

FIGURE 6.3
(a) Unfolding box; (b) Six-view third-angle projection.

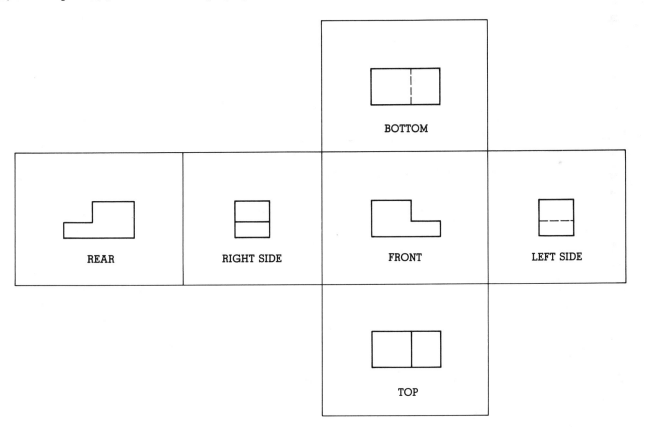

FIGURE 6.4
First-angle projection.

6.4 THE SELECTION AND PLACEMENT OF VIEWS

In a multiview drawing there is a certain order involved in the selection and placement of the several views of the object. View selection can be determined by examining the object to be drawn. If it is a complex part with several surfaces, holes, notches, and reliefs, it may require three or more views. If the object is a thin, flat piece or a round, spherical part,

FIGURE 6.5
Acceptable and unacceptable
relationships between front
view and top view.

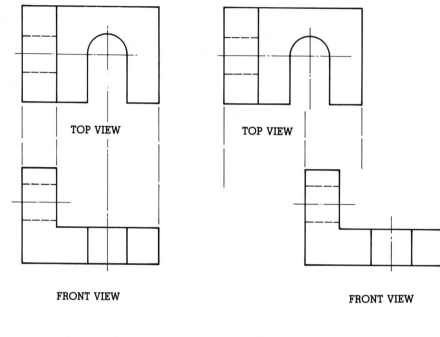

FIGURE 6.6
Acceptable and unacceptable
relationships between front
view and side view(s).

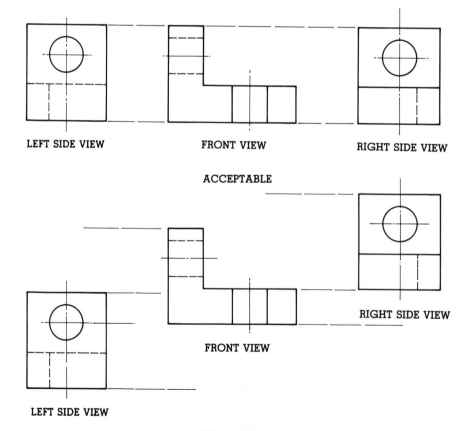

FIGURE 6.7
Acceptable and unacceptable relationships between the depths of the top view and the side view(s).

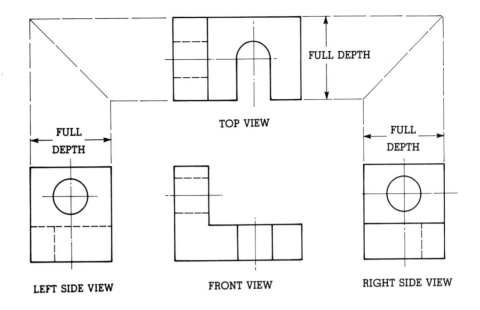

FULL DEPTH

TOP VIEW

FULL DEPTH

FULL DEPTH

LEFT SIDE VIEW   FRONT VIEW   RIGHT SIDE VIEW

ACCEPTABLE

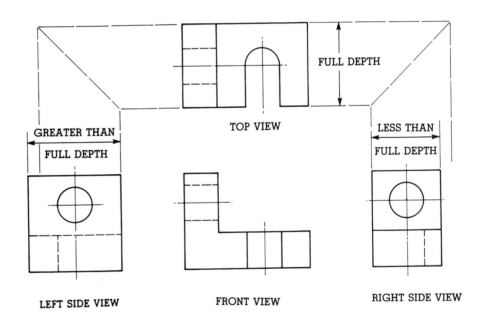

FULL DEPTH

TOP VIEW

GREATER THAN
FULL DEPTH

LESS THAN
FULL DEPTH

LEFT SIDE VIEW   FRONT VIEW   RIGHT SIDE VIEW

UNACCEPTABLE

only one or two views may be needed to clearly show its features. When trying to determine view selection, (1) attempt to draw those views that most clearly show the size and shape of the part, (2) select views that have the smallest numbers of hidden lines, and (3) make the front view the principal view—the view on which most dimensions can be placed.

View placement principles must be followed precisely if an acceptable multiview drawing is to be produced. The beginning drafter should possess the ability to place views correctly to be successful on the job. The important view placement principles include:

1. The front and top views must always be aligned vertically, with the top view above the front view, as illustrated in Figure 6.5.

FIGURE 6.8
Acceptable and unacceptable relationships between the lengths of the top view and front view.

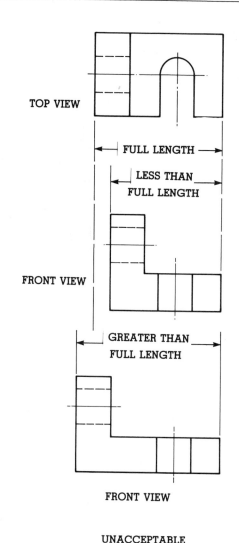

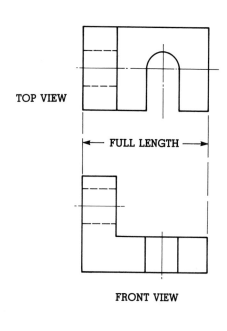

2. The front view and the side view(s) must always be aligned horizontally. The right side view must be drawn to the right of the front view, and the left side view must be drawn to the left of the front view, as shown in Figure 6.6.

3. The depth of the top view must always be the same as the depth of the side view(s), as represented in Figure 6.7.

4. The length of the top view must always be the same as the length of the front view, as portrayed in Figure 6.8.

5. The height of the side view(s) must always be the same as the height of the front view, as shown in Figure 6.9.

6. The bottom view must always be shown below and vertically aligned with the front view, as in Figure 6.10. It must also be of the same width as the front view.

6.5 THE ONE-VIEW DRAWING

The **one-view drawing** is acceptable for such items as cylinders, spheres, and square parts if important dimensions are properly indicated. Thin objects of uniform thickness, such as shims, gaskets, plates, and sheet-metal parts, may also be shown by single views as long as a note on the drawing provides important descriptors such as DIA, SQ, and THK (for diameter, square, and thick), as shown in Figure 6.11.

6.6 THE TWO-VIEW DRAWING

At times objects can be adequately described using two views. The **two-view drawing** may be arranged as any two adjacent views (views next to one another), as shown previously in Figure 6.3(b). Some

FIGURE 6.9
Acceptable and unacceptable relationships between the heights of the side view(s) and the front view.

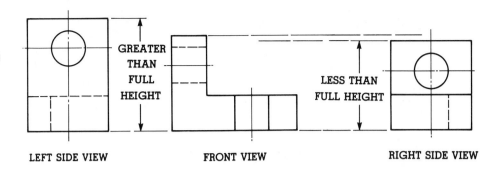

GREATER THAN FULL HEIGHT

LESS THAN FULL HEIGHT

LEFT SIDE VIEW

FRONT VIEW

RIGHT SIDE VIEW

UNACCEPTABLE

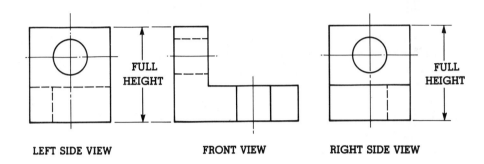

FULL HEIGHT

FULL HEIGHT

LEFT SIDE VIEW

FRONT VIEW

RIGHT SIDE VIEW

ACCEPTABLE

FIGURE 6.10
Acceptable and unacceptable relationships between bottom view and front view.

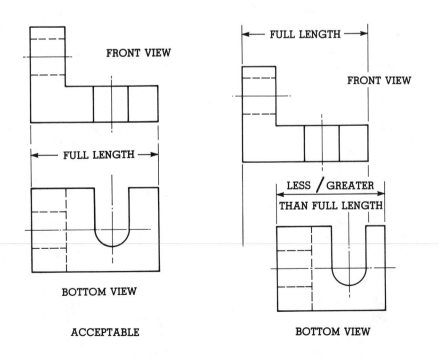

FRONT VIEW

FULL LENGTH

FRONT VIEW

FULL LENGTH

LESS / GREATER THAN FULL LENGTH

BOTTOM VIEW

ACCEPTABLE

BOTTOM VIEW

UNACCEPTABLE

drawings, such as those of cylindrical shapes, may consist of a top view and a front view or a front view and one side view. Addition of a third view would constitute unnecessary duplication of time, energy, paper, and information. When a two-view drawing is produced it is preferable to select views with the fewest hidden lines, as portrayed in Figure 6.12.

6.7 THE THREE-VIEW DRAWING

The **three-view drawing** may be arranged as any three adjacent views in the relationship identified in Figure 6.3(b). The most common views selected for the three-view drawing are the front, top, and right side views, but other arrangements are acceptable. If Figure 6.3(b) were divided into enough segments showing the object as a multiview drawing, the possibilities demonstrated in Figure 6.13 would result.

It can be seen in Figure 6.13(a) through (g) that some of the three-view drawings are acceptable while others are not because they show unnecessary hidden lines. Figure 6.13 only serves to illustrate the several different combinations of views that may be selected for three-view drawings. Figure 6.14 demonstrates examples of three-view drawings showing views that clearly outline the shape and size of the object.

6.8 THE PROJECTION OF POINTS, LINES, AND SURFACES

Producing the several views of an object is accomplished by taking measurements of certain points, lines, and surfaces from one view and projecting them to one or more other views. The projection of points, lines, and surfaces provides greater accuracy in alignment of views than does measuring each view separately with a scale or transferring them with dividers.

The use of a **miter line** provides a rapid and accurate method of projecting point, line, or surface measurements from one view to another, as demonstrated in Figure 6.15.

Note that the miter line is a single 45-degree angled line that originates at point O. D is the distance established by the drafter for space needed between views for dimensions, dimension and extension lines, and/or notes. It is a variable distance depending on space requirements.

In the production of a multiview drawing the drafter does not actually draw the surface of an object. Visible lines are drawn to indicate where one surface meets another surface. A line may indicate a

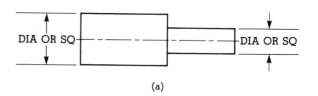

(a)

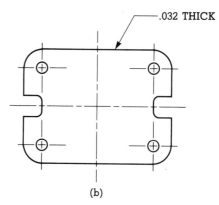

(b)

FIGURE 6.11
One-view drawing.

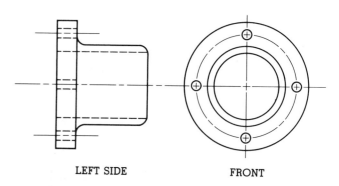

LEFT SIDE   FRONT

FIGURE 6.12
Two-view drawing.

surface in one view but may take the form of a point in another view. A line or surface that cannot be seen from a specific view because it is hidden or behind a surface is identified by a hidden or invisible (dashed) line. Figure 6.16 is a representation of a simple object which can be studied to learn the relationships among points, lines, and surfaces. A pictorial drawing of the object is also shown for ease of point, line, and surface visualization.

- **Points** A **point** is defined as something that has position, but no extension. It is usually indicated as a small crosshair, not as a dot. On a drawing it may be found at the intersection of two or more lines, at the termination point of a line, or at the corner of an object. When lines meet they form a

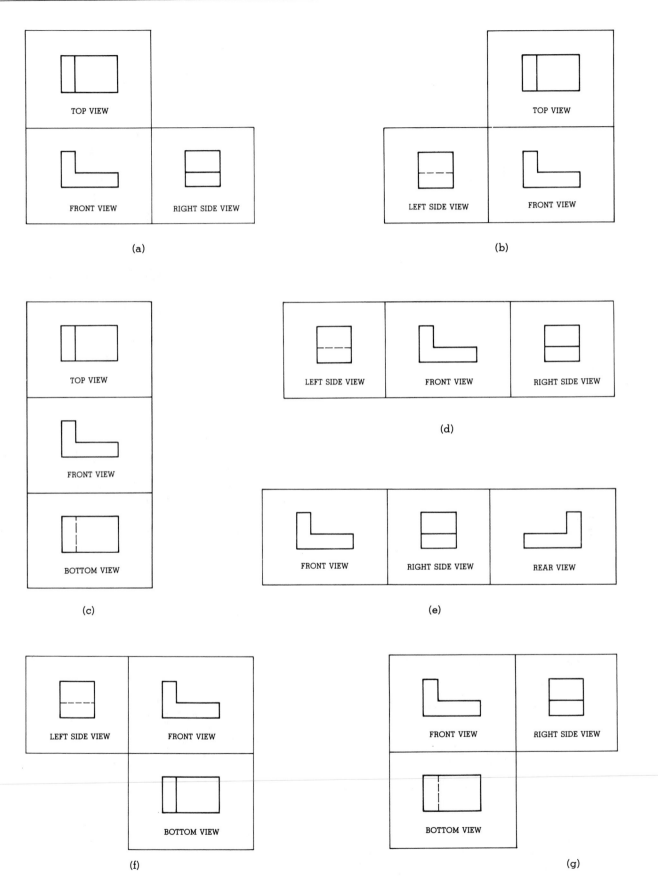

FIGURE 6.13
Examples of three-view drawings.

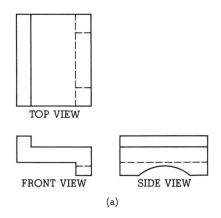

TOP VIEW

FRONT VIEW   SIDE VIEW

(a)

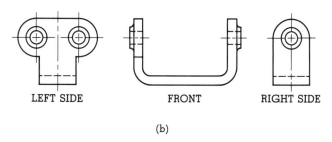

LEFT SIDE   FRONT   RIGHT SIDE

(b)

FIGURE 6.14
Three-view drawings clearly showing the shape and size of the object.

point. Figure 6.16 shows that: (1) lines 3 and 9 in the front view and line 7 in the side view form point X, as indicated in all three views and the pictorial view; and (2) lines 5 and 10 in the front view, 1 and

12 in the top view, and 6 in the side view form point Y.

- **Lines** A **straight line** was defined in Section 5.2. A curved line may be a variety of curved forms or arcs. There are four types of straight lines used in producing technical graphics: vertical, horizontal, inclined or slanted, and oblique. Each line is projected by locating its end point. In a multiview drawing, what takes the form of a surface in one view may become a line in an adjacent view. Figure 6.16 further demonstrates that: (1) surface A in the front view of the object becomes line 1 in the top view; (2) surfaces B, C, and D in the top view become lines 3, 5, and 4 in the front view and lines 7, 6, and 8 in the right side view (line 8 is shown as a hidden line corresponding to surface D); and (3) surfaces E and F in the right side view are shown as lines 9 and 10 in the front view and as lines 11 and 12 in the top view.

- **Surfaces** Most of the surface features on machine parts represent curved and/or plane areas. Examples of curved surfaces may be found on cylindrical shapes such as pistons, shafts, and holes. Plane surfaces are found on flat, cubic, and rectangular parts. Surfaces may be horizontal, vertical, inclined or slanted, curved, irregular, or oblique in shape. Figure 6.16 illustrates that: (1) surface A in the front view of the object is also surface A in the pictorial view, but is line 2 in the right side view and line 1 in the top view; and (2) vertical surfaces E and F in the right side view are shown as vertical lines 9 and 10 in the front view, vertical lines 11 and 12 in the top view, and surfaces E and F in the pictorial view.

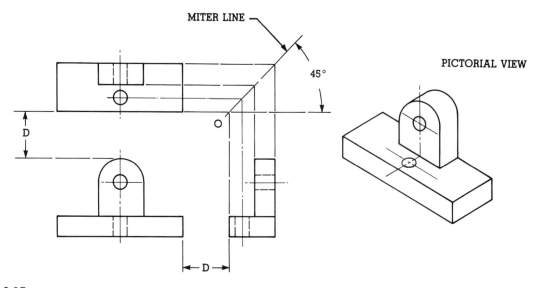

FIGURE 6.15
Projection of points, lines, and surfaces.

PICTORIAL VIEW

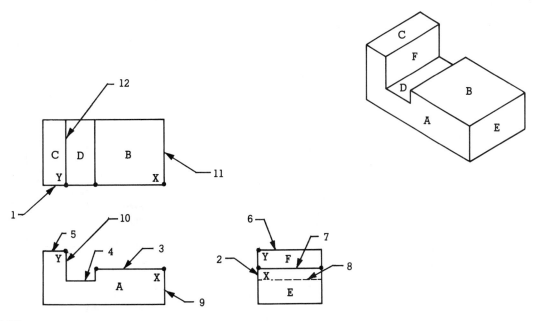

FIGURE 6.16
Relationships among points, lines, and surfaces.

6.9 THE PRECEDENCE OF LINES

To fully describe an object, a multiview drawing contains lines that represent edges, surfaces, intersections, size, limits, and points. In any given view there may be some parts of the object that are not clearly visible because they are covered by portions of the object that are closer to the eye of the observer. The areas of an object that are not clearly visible are therefore invisible and are represented by hidden lines. A hidden line consists of short dashes of uniform length with spaces between them, as illustrated in Figure 2.24.

The first lines produced on a multiview drawing are usually the center lines. Center lines are always light in weight and form the horizontal and vertical axes of circular features of cylinders, circles, spheres, and cones. Every circle or part of a circle shown must have its center represented by the intersection of two mutually perpendicular center lines. Center lines on a completed drawing consist of alternate long and short dashes, as shown in Figure 2.24.

On a multiview drawing, in any given view, there may be a conflict or a **coincidence of lines**. For example, a coincidence of lines occurs when an object line (visible line) projects to coincide with (or requires the same location as) a center line or a hidden line, or when a center line coincides with a cutting-plane line. Dimension and extension lines should always be positioned so as not to interfere or coincide with other lines. To make certain that there is a preferred standard "pecking order" for lines on a drawing—that is, a system for determining which lines hold preference or precedence over others—the **precedence of lines** was established. The following list defines the order of precedence of lines, which may also be referred to as the importance of lines:

1. Object or visible line
2. Hidden or invisible line
3. Cutting-plane line
4. Center line
5. Break line
6. Dimension and extension lines
7. Sectioning line

Figure 6.17 demonstrates coincidences of lines that result from separate features being of identical position, of identical size, or behind one another.

6.10 SUGGESTIONS FOR PRODUCING THE MULTIVIEW DRAWING

There is no definite order of steps that must be adhered to in producing a multiview drawing, but the following suggestions will assist the drafter in producing a quality document.

FIGURE 6.17
Precedence of lines.

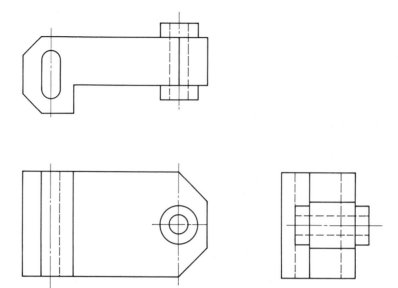

1. Mentally visualize or develop a rough freehand sketch of the object to be drawn.
2. Determine the number and combination of views required to best represent the object, remembering that the principal view should contain the majority of details and dimensions.
3. Select the appropriate size and type of drawing medium that will accommodate all the views, dimensions, notes, text, and specifications that will be needed.
4. Using light-weight lines and drafting equipment, draw the outlines of the object. Do this for all views, allowing space between views for dimensions. Determine if a miter line can be used to project the various views. Each hole, circle, or semicircle must have crossing center lines to indicate its center.
5. Add the necessary detail information to the object using light-weight lines. Draw all radii and circles first and connect all radii with visible lines, where required. It is better practice to draw a straight line to a radius or curve than to draw a radius or curve to a straight line.
6. When satisfied that the drawing is complete, view selection and placement are appropriate, sufficient space has been allowed for dimensions, notes, text, and specifications, and the precedence of lines has been maintained, "heavy up" all lines using the type and hardness of lead required to finalize the drawing.
7. At this point, make a copy of the drawing. Place the copy on the drafting board and add all the necessary data such as dimensions, notes, text, hole size, and thread information and specifica-

tions, until satisfied that the information is complete and properly located.
8. Transfer all information from the copy to the original drawing. Before doing this, in an effort to keep the drawing as clean and neat as possible, it may be desirable to place a sheet of plain paper between the hand and the drawing to keep perspiration from smudging the drawing.
9. Check the drawing for completeness and accuracy.

6.11 SUMMARY

The purpose of and need for the multiview drawing were discussed and the theory behind orthographic projection introduced. Detailed information on what the observer visualizes when viewing an object from different directions was presented for both third-angle and first-angle projection. The importance of selection and placement of views was stressed, and acceptable and unacceptable relationships between and among views were illustrated. Reference was made to one-view, two-view, and three-view drawings, showing examples of each and outlining their similarities and differences. Instructions and illustrations were provided for the projection of points, lines, and surfaces. Definitions of each of these terms were presented. Rules for establishing the precedence of lines were described, and the order of preference for various lines was identified. A step-by-step list of suggestions to aid the drafter in producing a multiview drawing was outlined. Practice exercises for this section are presented on pages 6A through 6J in the workbook.

6.12 **REVIEW EXERCISES**

1. What purpose does the multiview drawing serve?

2. Name the six views which are illustrated in an orthographic projection.

3. Identify three suggestions for determining view selection for a multiview drawing.

4. Describe six view placement principles which must be considered in producing a multiview drawing.

5. What is a miter line? Why is it useful?

6. Give four examples of parts for which a one-view drawing would be acceptable.

7. What are the most common view positions on a three-view drawing?

8. Define "point."

9. List three examples of objects with curved surfaces.

10. List three examples of objects that have plane surfaces.

11. Identify seven types of lines for which precedence has been established. List them in order of precedence.

12. What type of line is used to represent a surface or line that is not clearly visible to the observer?

13. Describe the function of the center line.

14. List five suggestions for producing an acceptable multiview drawing.

    a.

    b.

    c.

    d.

    e.

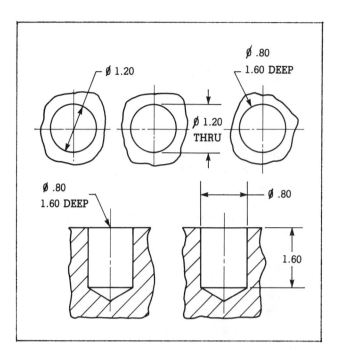

Section 7

Dimensioning

LEARNER OUTCOMES

The student will be able to:

- Identify the purpose of dimensions

- List the fundamental rules of dimensioning

- Define the most common terms used in dimensioning

- Explain the most frequently utilized dimensioning systems

- Perform appropriate practices for the dimensioning of features

- Apply dimensions to technical drawings in accordance with ANSI specification Y14.5M-1982

- Demonstrate learning through the successful completion of review and practice exercises

7.1 THE PURPOSE OF DIMENSIONS

Without dimensions, an object cannot be produced accurately. Dimensions on a drawing provide the producer of the object with the necessary information regarding the exact size and location of each feature of the part. A **dimension** is a numerical value expressed in appropriate units of measure. It is identified on a drawing, along with lines, symbols, and text, to define the geometrical characteristics of an object. Dimensions are used to complete the description of an object or a part. The ultimate user of the drawing must always be kept in mind when dimensions are being added to a drawing. For example, are dimensions being added to facilitate assembly, for machining purposes, to aid the mold maker, or for some other manufacturing or production purpose?

It is important that the drafter possess the skill and ability to determine, locate, and apply dimensions to drawings in an orderly, accepted manner. The most widely accepted standard for dimensioning is the American National Standards Institute (ANSI) specification Y14.5M-1982, "Dimensioning and Tolerancing for Engineering Drawings." Many medium-to-large-size industries, including government agencies, have changed their drafting practices to include this revised standard. The 1982 ANSI specification is an attempt to ensure interchangeability of parts produced by one manufacturer with apparently identical parts produced by another manufacturer. The dimensioning practices outlined in this section will reflect the various terms and symbols used in this specification.

7.2 FUNDAMENTAL RULES OF DIMENSIONING

For dimensions to define geometrical characteristics clearly and concisely, the drafter should follow a few basic rules in order to produce quality technical drawings. These rules will become obvious as one reads through this section. The rules include:

- Show enough dimensions so that the intended sizes and shapes can be determined without making calculations or assuming distances.
- State each dimension clearly, so that it can be interpreted in only one way.
- Show the dimensions between points, lines, or surfaces that have necessary and specific relations to each other or that control the locations of other components or mating parts.
- Dimension, extension, and leader lines shall not cross each other unless absolutely necessary.

- Select and arrange dimensions to avoid the accumulation of tolerances.
- Center lines, object lines, or extension lines should never be used as dimension lines.
- Show each dimension only once.
- Whenever possible, dimension each feature in the view where it appears.
- Dimensioning to hidden lines should be avoided, when possible.
- Place dimensions outside the outline of the part unless clarity is improved by doing otherwise.

Before applying dimensions, the drafter must first determine which view is the principal view of the object. The **principal view** is the one that most completely shows the object's characteristic shape. Show as many dimensions as practical in the principal view, but avoid overcrowding. All dimensions for surfaces that are shown in this profile should be given in this view. Figure 7.1 illustrates a three-view drawing with the principal view showing the most dimensions.

7.3 TYPES OF DIMENSIONING SYSTEMS

Several types of dimensioning systems are utilized in industry. They include the **unidirectional, aligned, tabular, arrowless**, and **chart drawing** types. The two in most prominent use are the unidirectional and aligned systems of dimensioning and will be covered here.

7.3.1 The Unidirectional System

The unidirectional system is probably the system most commonly used for mechanical technical draft-

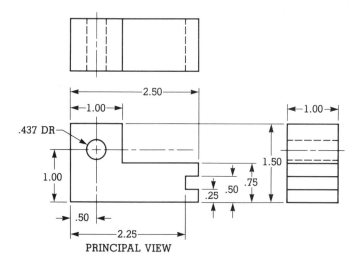

FIGURE 7.1
Three-view drawing.

ing purposes. It is the easiest system to read because it requires that all numerals, figures, and notes be lettered horizontally and be read parallel to the bottom edge of the drawing medium. Figure 7.2 shows an object which is dimensioned using this method.

7.3.2 The Aligned System

The aligned system places the numerals in alignment with dimension lines so that they must be read from the bottom for horizontal dimensions and from the right side for vertical dimensions. Figure 7.3 demonstrates acceptable and unacceptable dimensioning practices for the aligned system.

7.4 DEFINITIONS OF DIMENSIONING TERMS

Several technical terms that are considered important to the topic are identified below. It is necessary that the drafter become familiar with their meanings and usages.

- **Basic dimension** A numerical value used to describe the theoretical exact size, profile, orientation, or location of a feature or a datum target. It is the basis from which permissible variations are established by tolerances on other dimensions, in notes, or in feature control frames.
- **True position** The exact location of a feature established by basic dimensions.
- **Reference dimension** A dimension, usually without tolerance, used for information purposes only. It is considered to be auxiliary information and does not govern production or inspection operations. A reference dimension is a repeat of a dimension or is derived from other values shown on the drawing or on related drawings. The method for identifying a reference dimension is to enclose it in parentheses, as shown in Figure 7.4.

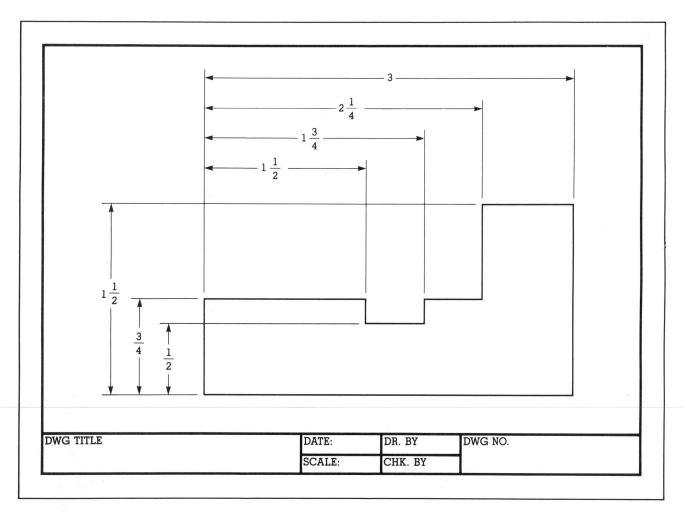

FIGURE 7.2
Unidirectional dimensioning practices.

FIGURE 7.3
Acceptable and unacceptable
aligned drafting practices.

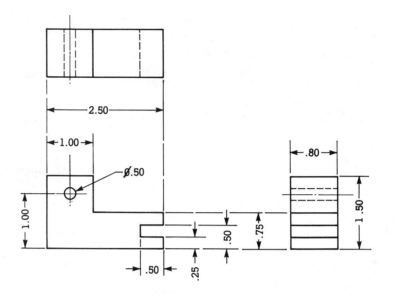

ACCEPTABLE

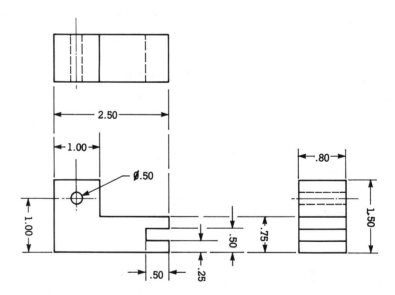

UNACCEPTABLE

- **Datum** A theoretically exact point, axis, or plane derived from the true geometric counterpart of a specified datum feature. A datum is the origin from which the locations or geometric characteristics of features of a part are established. An illustration of a datum is shown in Figure 7.5, where the datum is point O at the intersection of the x and y axes (shown as A and B, respectively).

- **Datum target** A specified point, line, or area on a part which is used to specify a datum.
- **Feature** A general term which refers to a physical attribute of a part, such as a surface, hole, or slot. Figure 7.6 demonstrates such features.
- **Feature of size** A part feature that has thickness or round characteristics. These may be cylindrical shapes, such as shafts and holes, or rectangular

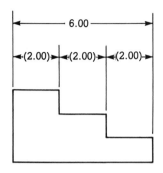

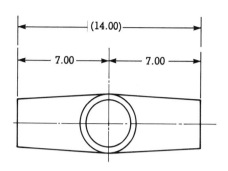

(a) INTERMEDIATE REFERENCE DIMENSION

(b) OVERALL REFERENCE DIMENSION

FIGURE 7.4
Methods for identifying reference dimensions.

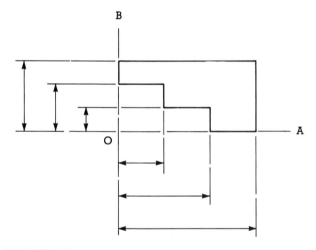

FIGURE 7.5
Illustration of a datum.

or flat parts where two parallel or flat surfaces form a single feature. Examples of features of size are depicted in Figure 7.7.

- **Datum feature** An actual feature of a part that is used to establish a datum. A datum feature is selected on the basis of its geometric relationship to the toleranced feature and the requirements of the design.
- **Actual size** The measured size.

7.5 TYPES OF DIMENSIONS

As a rule, most drawings are dimensioned by use of decimal units. Exceptions include drawings produced for architectural, civil, and structural steel design and fabrication. In addition, because several large industries are multinational in their organizational make-up, many drawings of machined parts are produced using metric units. So drawings produced in the United States may use decimal inch dimensions, millimeter dimensions, or both types of dimensions simultaneously.

7.5.1 The Decimal Inch Dimension

In this system of dimensioning, the term **decimal inch** designates the use of subdivisions or multiples of an inch in decimal notation to whatever numbers of decimal places are required. These designations may include decimal units as subdivisions or multiples of an inch, square inch, or cubic inch. All dimensions are expressed as decimals except for angular dimensions, which continue to be expressed both in degrees, minutes, and seconds, and as decimals, as demonstrated in Figure 7.8.

The basis for the decimal inch system is the use of a two-place decimal utilizing an increment of .02 inch or multiples of .02, such as .04, .24, 3.20, etc.

FIGURE 7.6
Part features.

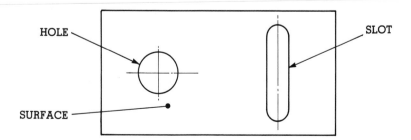

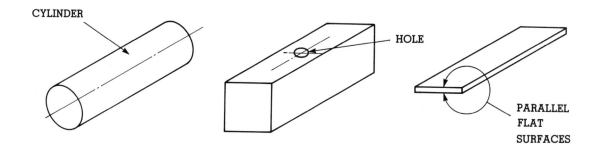

FIGURE 7.7
Features of size.

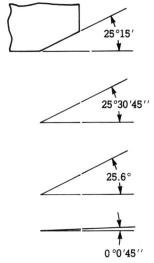

FIGURE 7.8
Angular units.

For values less than one, zeroes do not precede the decimal point. Figure 7.9 illustrates examples of such dimensions.

7.5.2 The Millimeter Dimension

Several conventions have been adopted for utilizing the **millimeter dimension** that were derived from the **International System of Units (SI)** and are expected to have wide use in the United States. The following conventions are normally observed when specifying millimeter dimensions on drawings and are depicted in Figure 7.10.

1. Where the dimension is less than one millimeter, a zero precedes the decimal point.

2. Where the dimension is a whole number, neither the decimal point nor a zero is shown.

3. Where the dimension exceeds a whole number by a decimal fraction of one millimeter, the last

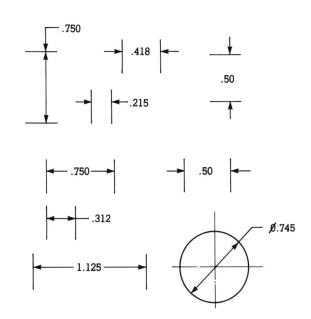

FIGURE 7.9
Decimal inch dimensions.

digit to the right of the decimal point is not followed by a zero.

4. Neither commas nor spaces shall be used to separate digits into groups in specifying millimeter dimensions on drawings.

7.6 THE APPLICATION OF DIMENSIONS

Before the drafter begins a drawing of an object, consideration should be given to how much space the views, dimensions, and notes will consume. Dimensions should always be legible, placed at a uniform distance from the object, uncrowded, and placed between views whenever possible. Dimensions are applied using dimension lines, extension

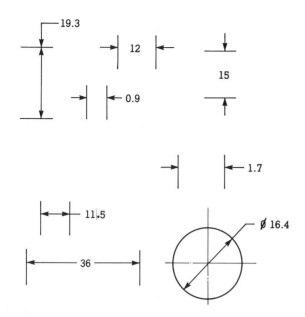

FIGURE 7.10
Conventions for millimeter dimensions.

lines, or leaders from dimensions, local notes, or specifications directed to the appropriate feature. General notes or text are used to convey additional information.

7.6.1 The Dimension Line

The **dimension line**, with terminations called "arrowheads," shows the direction and extent of a dimension. Numerals indicate the number of units of a measurement. Preferably, dimension lines should be broken for insertion of numerals, as shown in Figure 7.11. When a horizontal dimension line is not broken, numerals are placed above and parallel to the dimension line.

Dimension lines are aligned, if possible, and grouped for easier readability and uniform appearance as illustrated in Figure 7.12.

Dimension lines should always be drawn parallel to the direction of measurement. The space between the first dimension line and the object outline should not be less than .40 inch. The space between succeeding parallel dimension lines should not be less than .25 inch. Figure 7.13 depicts this recommendation. In addition, there should be a visible gap between the object and an extension line.

When there are several parallel dimension lines and space is restricted, it is better to stagger columns of dimensions to eliminate the possible interference of numerals. Figure 7.14 portrays staggered dimensions.

There are certain lines that should never be used as dimension lines. These include center lines, extension lines, phantom lines, lines that are part of

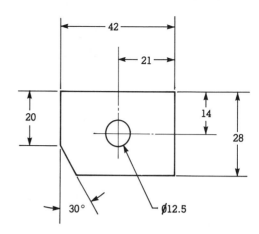

FIGURE 7.11
Dimension lines.

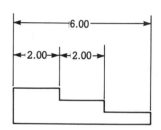

FIGURE 7.12
Grouping of dimensions.

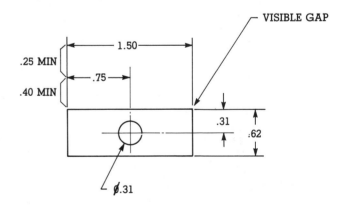

FIGURE 7.13
Spacing of dimensions.

the outline of the object, or continuations of any of these lines. In addition, crossing of dimension lines should be avoided. When dictated by unavoidable circumstances, dimension lines are unbroken.

7.6.2 The Extension Line

The **extension line**, which is also referred to as the **projection line**, is used to indicate the extension of a surface or point to a location outside the part outline. As indicated in Figure 7.13, an extension line

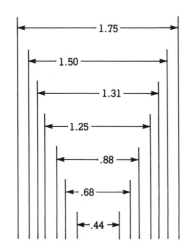

FIGURE 7.14
Staggered dimensions.

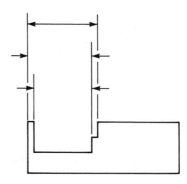

FIGURE 7.16
Breaks in extension lines

begins with a short visible gap from the outline of the part and extends beyond the outermost related dimension line. Extension lines are usually drawn perpendicular to dimension lines. Where space is restricted, they may be drawn at an oblique angle, to clearly illustrate the application. When oblique extension lines are used, the dimension line is shown in the direction in which the dimension applies. Typical extension lines are represented in Figure 7.15.

When possible, extension lines should not cross one another, nor should they cross dimension lines. To minimize such crossings, the shortest dimension line is shown nearest the object's outline. When it is necessary for extension lines to cross other extension lines, dimension lines, or lines showing part features, they are not broken. Breaks in extension lines are used when the lines cross arrowheads or dimension lines close to arrowheads, as demonstrated in Figure 7.16.

7.6.3 The Leader

The **leader** is a line which is used to direct a dimension, note, or symbol to the intended location on the drawing. It takes the form of an inclined straight line except for a short horizontal portion extending to mid-height of the first or last letter or digit of a note or dimension, as indicated in Figure 7.17. Two or more adjacent leaders are drawn parallel.

The leader normally terminates in an arrowhead touching a line. When the leader makes reference to a surface by ending within the outline of that surface, the leader will terminate in a dot at least .06 inch in diameter. Examples are shown in Figure 7.18.

When a leader is directed to an arc or a circle, its direction should be radial. That is, the leader should point to the center of the arc or circle, as illustrated in Figure 7.19.

To avoid confusion, the following should be avoided, if at all possible, when drawing leaders:

1. Crossing of other leaders
2. Long leaders
3. Leaders drawn in a horizontal or vertical direction
4. Leaders adjacent to dimension lines, extension lines, or section lines

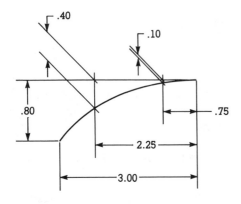

FIGURE 7.15
Extension lines.

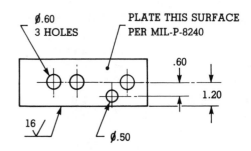

FIGURE 7.17
Leaders

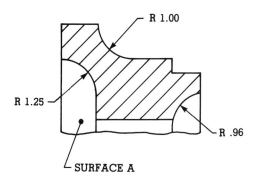

FIGURE 7.18
Leader-directed dimensions.

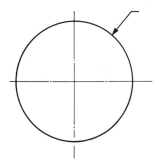

FIGURE 7.19
Correct positioning of a leader.

7.6.4 Dimensions Not to Scale

When it is necessary or desirable to indicate that a particular feature or dimension is not drawn to scale, the dimension in question should be underlined with a straight thick line, as depicted in Figure 7.20.

7.7 THE DIMENSIONING OF FEATURES

Various characteristics and features of objects require special or unique methods of dimensioning. The practices that follow are considered to be important enough for the drafter to become familiar with them.

7.7.1 Diameters

In utilizing the ANSI 1982 specification, the **diameter symbol** precedes all diametral values. When the diameter of a spherical feature is required, the diametral value is preceded by the **spherical diameter symbol, S**. Refer to Figures 8.10 and 8.11 for geometric characteristic symbols for geometric tolerancing. When the diameters of a number of concentric cylindrical features are required, such diameters should be dimensioned in a longitudinal view, if

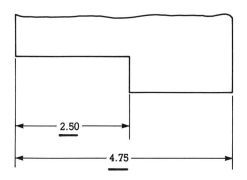

FIGURE 7.20
Dimensions not to scale.

practical. Figure 7.21 portrays the dimensioning of diameters.

7.7.2 Radii

When a radius value is specified on a drawing it is preceded by the **radius symbol, R**. A radius dimension line utilizes one arrowhead, at the arc end. An arrowhead is never shown at the radius center. When location of the center is important and space permits, a dimension line is drawn from the radius center and the arrowhead touches the arc. In this case the dimension is placed between the arrowhead and the center.

When space is limited, the dimension line is extended through the radius center. If it is inconvenient to place the arrowhead between the radius center and the arc, it may be placed outside the arc with a leader. When the center of a radius is not dimensionally located, the center shall not be indicated. The above examples are illustrated in Figure 7.22.

When the center of a radius is located by dimensions, a crosshair is drawn at the center. Extension and dimension lines are used to locate the center. If the location of the center is not important, the drawing should clearly show that the location of the arc is controlled by other dimensioned features such as tangent surfaces. Figure 7.23 demonstrates di-

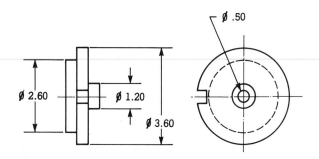

FIGURE 7.21
Dimensioning of diameters.

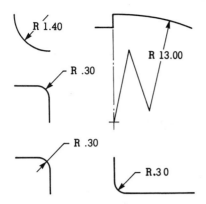

FIGURE 7.22
Dimensioning of radii.

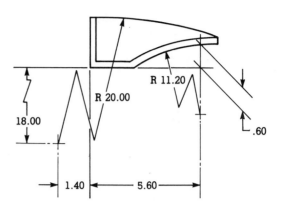

FIGURE 7.24
Dimensioning of external radii.

mensioning practices for radii with located and un-located centers.

If the center of a radius is located outside the drawing or interferes with another view, the radius line may be shortened. The portion of the dimension line extending from the arrowhead is radial relative to the arc. When the radius dimension line is shortened and the center is located by coordinate dimensions, the dimension line which locates the center is also shortened. Figure 7.24 identifies methods for dimensioning radii located external to the drawing.

7.7.3 Chords, Arcs, and Angles

The dimensioning of chords, arcs, and angles is shown in Figure 7.25. For a chord, the dimension line is perpendicular to the extension lines and parallel with the chord. For an arc, the dimension line follows the contour of the arc curve, and the extension lines are either vertical or horizontal. In addition, the **arc symbol** ⌒ is placed above the dimension. For dimensioning of an angle, the extension lines project from the sides forming the angle and the dimension line forms an arc.

7.7.4 Rounded Ends

Overall dimensions are used on parts having rounded ends. For fully rounded ends, the radii are indicated but not dimensioned, as illustrated in Figure 7.26(a). For parts with partially rounded ends, the radii are dimensioned, as depicted in Figure 7.26(b).

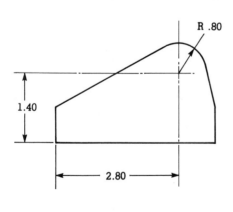

(a) LOCATED CENTER

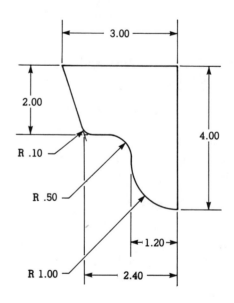

(b) UNLOCATED CENTER

FIGURE 7.23
Radii with located and unlocated centers.

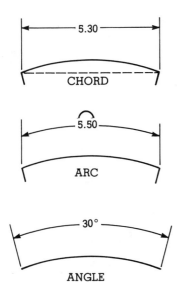

FIGURE 7.25
Dimensioning of chords, arcs, and angles.

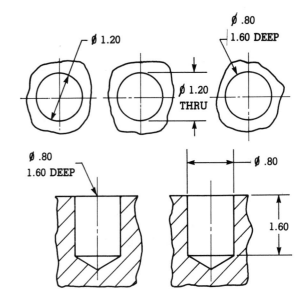

FIGURE 7.27
Dimensioning practices for round holes.

7.7.5 Holes

There are many different types of holes to be dimensioned. Some are round, others are not. Some are slanted, others require different depths and contours. Holes are one of the most commonly dimensioned types of features on a drawing and need to be described and indicated clearly by the drafter.

7.7.5.1 The Round Hole

The **round hole** is dimensioned as portrayed in Figure 7.27. When it is not clear whether or not the hole goes completely through the object, the abbreviation THRU follows the dimension. When a hole does not extend completely through the object it is called a **blind hole**. The depth dimension of a blind hole is the depth of the full diameter from the outer surface of the part. If a blind hole is also counterdrilled or counterbored, the depth dimension is also measured from the outer surface.

7.7.5.2 The Slotted Hole

The **slotted hole**, or **slot**, as it is often called, is usually dimensioned by length and width in one of the three methods illustrated in Figure 7.28. In Figure 7.28(a), the slotted hole is dimensioned using a longitudinal center line and a width dimension. In Figure 7.28(b), the length and width are indicated by a local note. Figure 7.28(c) indicates overall length and width by use of dimension lines. In each case the end radii are shown but not dimensioned.

7.7.5.3 The Counterbored Hole

The major purpose of the **counterbored hole** is to provide a flat, flush seat for a fastener. Counterbored holes are often provided in castings that have rough surfaces on which smooth areas are required for seating of screws. The appropriate notation for a counterbored hole is CBORE, as demonstrated in

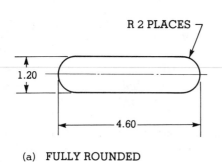

(a) FULLY ROUNDED

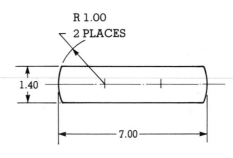

(b) PARTIALLY ROUNDED

FIGURE 7.26
Dimensioning of fully and partially rounded ends.

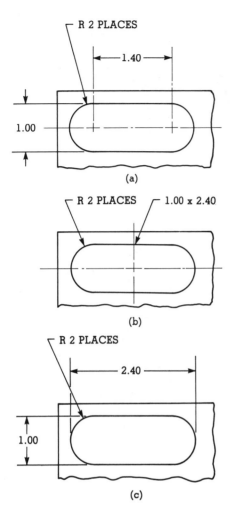

FIGURE 7.28
Dimensioning of slotted holes.

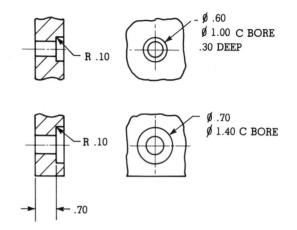

FIGURE 7.29
Dimensioning of counterbored holes.

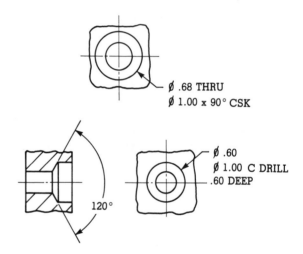

FIGURE 7.30
Dimensioning of countersunk and counterdrilled holes.

Figure 7.29. A counterbored hole is specified by a note giving the diameter, depth, and inside corner radius. If the thickness of the remaining material is of some importance, this thickness, rather than the depth, is dimensioned.

7.7.5.4 The Countersunk Hole and the Counterdrilled Hole

The purpose of a **countersunk hole** or a **counterdrilled hole** is to obtain a flush or below-surface mounting condition for a screw or a rivet. Countersinking is similar to counterboring. The difference (see Figure 7.30) is that the countersink is at an angle. The designated symbol is CSK. For a countersunk hole, the diameter and the included angle of the countersink are specified. For a counterdrilled hole, whose designation is CDRILL, the diameter, depth, and included angle of the counterdrill are indicated. As stated earlier for round holes, the depth dimen-

sion is the depth of the full diameter of the counterdrill from the outer surface of the object.

7.7.5.5 The Spotface

Machining of a **spotface** is often performed on a casting to clean up the surface or to provide an accurately machined area for a washer or the head of a fastener. On a drawing, the diameter of the spotface is always specified. Either the depth or the remaining material thickness may be specified, as shown in Figure 7.31. The spotface, whose symbol is SF, may also be specified by a note, and in this case it need not be delineated on the drawing. If the spotface is merely used to clean up the surface, no depth or remaining material thickness is specified.

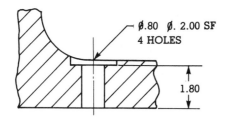

FIGURE 7.31
Dimensioning of spotfaced holes.

7.7.5.6 The Chamfer

The **chamfer** is produced to eliminate sharp edges or burrs on machined objects. There are basically two types of chamfers: external and internal. The **external chamfer** is dimensioned using an angle and a linear dimension, or with two linear dimensions. When an angle and a linear dimension are used, the linear dimension is the distance from the indicated surface of the object to the start of the chamfer, as shown in Figure 7.32(b). A note may be used to specify 45-degree chamfers—but only 45-degree chamfers, because both the radial and longitudinal values are the same, as illustrated in Figure 7.32(a).

For the **internal chamfer** on the edge of a round hole, the same practice is used as for the external chamfer except where dimensional control of the chamfer is required. Two examples are illustrated in Figure 7.33.

When chamfers are required for surfaces intersecting at other than 90 degrees, the practice depicted in Figure 7.34 is recommended.

7.7.5.7 The Keyseat

The **keyseat** is a seat for a key. It is a slot in a shaft or a hub, or both, to hold a key. The **key**, which usually takes the form of a square in cross section, is used to

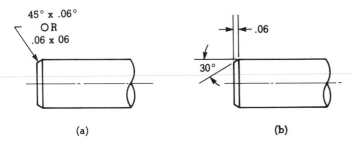

FIGURE 7.32
Dimensioning of external chamfers.

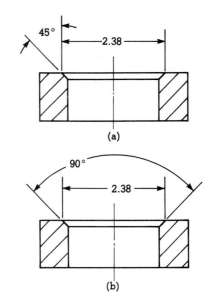

FIGURE 7.33
Dimensioning of internal chamfers.

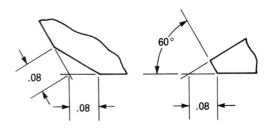

FIGURE 7.34
Dimensioning of chamfers for surfaces at other than 90 degrees.

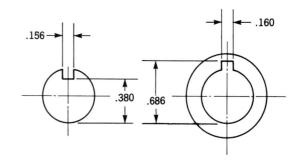

FIGURE 7.35
Dimensioning of keyseats.

ensure that both the shaft and its mating part will rotate together. The keyseat is dimensioned by width, depth, location, and length, as illustrated in Figure 7.35. The depth of the keyseat is normally dimensioned from the opposite side of the shaft or hole.

7.8 **SUMMARY**

This section brought under discussion the broad topic of dimensioning. The purpose and fundamental rules of dimensioning were presented, and types of dimensioning systems were detailed. Several dimensioning terms were defined, and significant types of dimensions were covered. Numerous aspects of the application of dimensions were detailed and interpreted, and several examples were presented. Appropriate dimensioning practices for some 15 different features were discussed and an example of each was provided. Practice exercises for this section are presented on pages 7A through 7Q in the workbook.

7.9 REVIEW EXERCISES

1. What purpose does the dimension serve?

2. What is the most accepted standard or specification used for dimensioning today?

3. List five fundamental rules for dimensioning.

4. In orthographic projection, in which view should most of the dimensions be placed?

5. Identify the two most prominently used dimensioning systems. Define each.

6. Define the term "basic dimension".

7. What is a datum?

8. Give examples of part "features".

9. What is the purpose of a leader?

10. How is a "not to scale" dimension shown?

11. How and when is an extension line used?

12. Describe the decimal inch system of dimensioning.

13. From what was the millimeter system of dimensioning derived?

14. Identify three types of lines that should never be used as dimension lines.

15. What is the purpose of a countersunk or counterbored hole?

16. What is a chamfer?

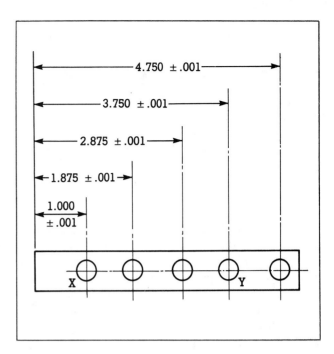

Section 8

Tolerancing

LEARNER OUTCOMES

The student will be able to:

- Define tolerancing terms
- Explain the emergence of tolerancing
- State the purpose of tolerancing
- Explain tolerancing methods
- Identify geometric characteristic symbols
- Apply geometric tolerance principles to the technology of drafting
- Demonstrate learning through the successful completion of review and practice exercises

8.1 THE NEED FOR TOLERANCING

The advent of mass production and the demand for high quality have led to a need for interchangeability of parts and assemblies. Parts for machines and structures must be produced to some acceptable size or standard so that when they are assembled they will fit and work properly. From the effort to achieve high quality and assurance that a part made at one production facility will be identical to one produced at another facility, the concept of **tolerancing** emerged. When there is a string or chain of dimensions on a part, the positions of the dimensioned features may vary as a result of the overall accumulation of individual allowances.

It is virtually impossible to produce parts to exact size. So that parts will fit and function as required while still being economical to manufacture, producers are given an appropriate variation, or margin of error, on each dimension. This allowable variation is called a **tolerance**.

8.2 THE PURPOSE OF TOLERANCING

The purpose of tolerancing is to control the dimensions of any two or more mating parts so that they will be interchangeable. Interchangeability implies that a part can change places with another part of the same size, form, and dimensions.

An example of a detail part whose dimensions need to be controlled is the pin shown in Figure 8.1.

The diameter of the shank of the pin can be as large as .4955 inch or as small as .4945 inch, or can be of any value between the two. If the diameter is acceptable it is said to be **within tolerance**. If the diameter falls above or below the **limit**, it is said to be **out of tolerance**. In this case the tolerance, which is the difference between the two numbers, is .0010 inch (read "ten ten-thousandths").

8.3 TOLERANCING TERMS

Besides the dimensioning terms referred to in subsection 7.4, the definitions listed below are related more specifically to tolerancing. They include:

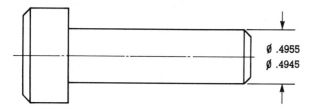

FIGURE 8.1
Pin with tolerance.

- **Nominal size** The approximate size of the part, written in its fractional form. The nominal size of the pin in Figure 8.1 is 1/2 inch diameter since both decimal dimensions are close to 1/2 inch.
- **Basic size** The exact theoretical size of the part, presented in decimal form. This is the starting point for working out the tolerance for a dimension. The basic size of the shank of the pin in Figure 8.1 is .5000 inch.
- **Limits** or **limit dimensions** The largest and smallest acceptable dimensions. For the pin shank in Figure 8.1:

    .4955 is the high limit (maximum value)
    .4945 is the low limit (minimum value)

- **Mating parts** Parts which will be fastened or assembled together, such as a shaft in a hole, a screw in a tapped hole, or a key in a slot.
- **Fit** The looseness or tightness relationship between two mating parts when assembled.
- **Allowance** The fit between two mating parts. There are two types of allowances: interference and clearance.
- **Interference allowance** Interference allowance occurs when a shaft is larger than the hole, causing the shaft to be pressed or forced into the hole.
- **Clearance allowance** A type of fit which provides clearance or space between mating parts. It is the opposite of an interference allowance.
- **Regardless of feature size (RFS or S)** A condition wherein the tolerance of position or form must be met regardless of feature or datum size.
- **Tolerance** The total amount by which a specific dimension is permitted to vary. The tolerance is the difference between the maximum and minimum limits, as illustrated in Figure 8.2.
- **Unilateral tolerance** A tolerance in which variation is permitted in one (**uni**) direction from the specified dimension. Examples are:

$$.875 \begin{array}{c} + .000 \\ - .005 \end{array} \text{ and } .750 \begin{array}{c} + .003 \\ - .000 \end{array}$$

- **Bilateral tolerance** A tolerance in which variation is permitted in both (**bi**) directions from the specified dimension. For example:

$$.375 \pm .005 \text{ and } .250 \pm .003$$

- **Geometric tolerance** The general term applied to the category of tolerances used to control form, profile, orientation, location, and runout.
- **Limits of size** Limits that describe the extent within which variations of geometric form, as well as size, are allowed. They are the specified maximum and minimum sizes, as portrayed in the dimensions of the hole in Figure 8.3.

FIGURE 8.2
Tolerance example.

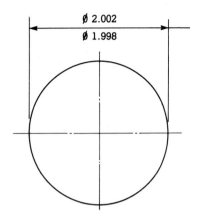

Ø 2.002 — MAXIMUM LIMIT
Ø 1.998 — MINIMUM LIMIT

2.002
1.998
──────
DIFFERENCE = .004 = TOLERANCE

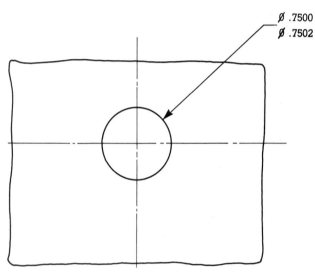

Ø .7500
Ø .7502

FIGURE 8.3
Limits of size.

- **Maximum material condition (MMC)** The condition wherein a feature of size contains the maximum amount of material within the stated limits of size—for example, the minimum hole diameter and the maximum shaft diameter. A pictorial example of MMC is shown in Figure 8.4.
- **Least material condition (LMC)** The condition wherein a feature of size contains the least amount of material within the stated limits of size—for example, the maximum hole diameter and the minimum shaft diameter. In Figure 8.4, the LMC for the shaft would be:

    LMC = .500 − .005 = .495 for the shaft

and the LMC for the hole would be:

    LMC = .500 + .003 = .503 for the hole

8.4 TOLERANCING METHODS

When applying a tolerance directly to a dimension, there is a certain order that the drafter should follow.

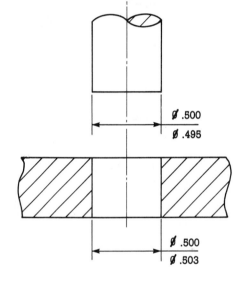

Ø .500
Ø .495

Ø .500
Ø .503

MAXIMUM SHAFT DIAMETER = .500 = MMC FOR SHAFT

MINIMUM HOLE DIAMETER = .500 = MMC FOR HOLE

FIGURE 8.4
Maximum material condition (MMC).

1. The high limit (maximum value) is placed above the low limit (minimum value) for shafts, keys, and the like, for outside dimensions of an object, as presented in Figure 8.5(a) and (b). For holes and slots the low limit (minimum value) is placed above the high limit (maximum value), as shown in Figure 8.5(c) and (d).
2. In placing a dimension on a drawing, the dimension itself appears first followed by a plus-or-minus expression of tolerance, as indicated in Figure 8.6(a) and (b).

8.5 TOLERANCE EXPRESSION

For adding tolerance limits to a dimension, the plus-or-minus tolerance and the dimension itself are ex-

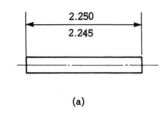

(a)

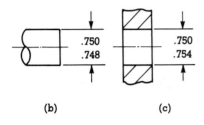

(b)   (c)

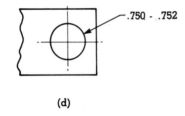

(d)

FIGURE 8.5
Tolerancing limits.

pressed in the same form and to the same number of places. Figure 8.7 clarifies this point by presenting acceptable and unacceptable practices.

8.6 INTERPRETATION OF LIMITS

All limits are considered to be absolute—that is, accepted without question. Therefore, dimensional limits, regardless of the number of decimal places, are identified as if they were continued with zeros. For example:

1.22	means	1.220 0
1.20		1.200 0

1.202	means	1.2020 0
1.200		1.2000 0

8.6.1 Single Limits

Minimum and maximum dimensions control the limits. When this requirement occurs, the abbreviation MIN or MAX is placed after a numeral where other elements of the design determine the other unspeci-

(a) UNILATERAL TOLERANCING

(b) BILATERAL TOLERANCING

FIGURE 8.6
Plus-or-minus tolerancing.

ACCEPTABLE	UNACCEPTABLE
.750 .748	.75 .748
.500 ± .005	.50 ± .005
6 FT 2.25 ± .25	6 FT 2.25 ± 0 FT .25

FIGURE 8.7
Acceptable and unacceptable expressions of tolerance.

fied limit. Part features such as depths of holes, lengths of threads, corner radii, and chamfers may be limited in this manner. **Single limits** are used where (a) the intent of the limit is clear, and (b) the unspecified limit is relatively unimportant. An example of how a single limit is used is presented in Figure 8.8.

8.7 THE ACCUMULATION OF TOLERANCES

When the location of a surface in a given direction is affected by more than one tolerance, the tolerances are considered to be **cumulative**. If the accumulation of tolerances has a negative effect on the part—that is, renders it unacceptable for quality—then the

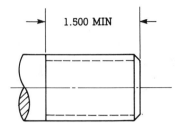

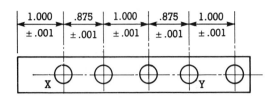

(a) CHAIN DIMENSIONING - GREATEST TOLERANCE
ACCUMULATION BETWEEN X AND Y

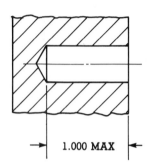

FIGURE 8.8
Applying MIN and MAX limits.

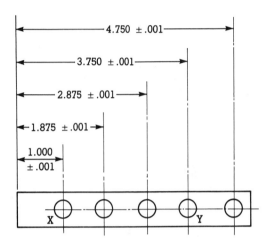

(b) DATUM DIMENSIONING - LESSER TOLERANCE
ACCUMULATION BETWEEN X AND Y

drafter needs to determine which type of dimensioning would be appropriate so that acceptable parts can be produced. Figure 8.9 compares the tolerance accumulations resulting from three different methods of dimensioning.

Chain dimensioning, shown in Figure 8.9(a), results in the greatest tolerance accumulation, as illustrated by the ±.003 variation between holes X and Y.

Datum dimensioning, depicted in Figure 8.9(b), reduces the total tolerance accumulation between holes X and Y to ±.002.

Direct dimensioning, shown in Figure 8.9(c), results in the least tolerance accumulation, as illustrated by the ±.001 variation between holes X and Y.

There is no single best method of dimensioning and tolerancing. The designer and drafter need to review the requirements of the project and base dimensioning or tolerancing methods in order to produce the highest quality parts for the most economical cost.

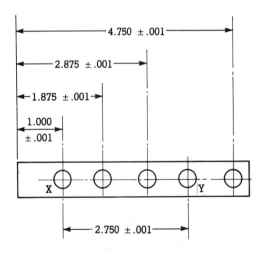

(c) DIRECT DIMENSIONING - LEAST TOLERANCE
ACCUMULATION BETWEEN X AND Y

FIGURE 8.9
Accumulation of tolerances.

8.8 SYMBOLOGY

Symbols are used to specify geometric characteristics on drawings involving technical graphics. Symbols are preferred to notes and text because they convey the intended meaning in a precise and direct form. In addition, their use tends to reduce errors due to improperly worded notes.

FIGURE 8.10
Geometric characteristic
symbols.

			CHARACTERISTIC	
INDIVIDUAL FEATURES	FORM TOLERANCE		STRAIGHTNESS	—
			FLATNESS	▱
			ROUNDNESS	○
			CYLINDRICITY	⌭
INDIVIDUAL OR RELATED FEATURES			PROFILE OF A LINE	⌒
			PROFILE OF A SURFACE	⌓
RELATED FEATURES			ANGULARITY	∠
			PERPENDICULARITY	⊥
			PARALLELISM	//
	LOCATION TOLERANCE		POSITION	⌖
			CONCENTRICITY	◎
			SYMMETRY	≡
	RUNOUT TOLERANCE		CIRCULAR	↗
			TOTAL	↗

8.8.1 The Geometric Characteristic Symbol

The symbols which are used when denoting geometric characteristics are presented in Figure 8.10. Other symbols used with feature control symbols are identified in Figure 8.11.

8.8.2 The Datum Identifying Symbol

The **datum identifying symbol** consists of a frame containing a **datum reference letter** preceded and followed by dashes, as shown in Figure 8.12. The symbol is associated with a datum feature by one of the methods outlined in Figure 8.22.

Any letter of the alphabet, with the exception of I, O, and Q, may be used as a datum reference letter. The symbol indicates the datum on the drawing, and the letter identifies the datum. When more than one datum is used on a drawing, each one is identified with a different datum letter.

TERM	ABBREVIATION	SYMBOL
MAXIMUM MATERIAL CONDITION	MMC	Ⓜ
REGARDLESS OF FEATURE SIZE	RFS	Ⓢ
DIAMETER	DIA	⌀
PROJECTED TOLERANCE ZONE	TOL ZONE PROJ	Ⓟ
REFERENCE	REF	(1.250)
BASIC	BSC	3.875

FIGURE 8.11
Other symbols.

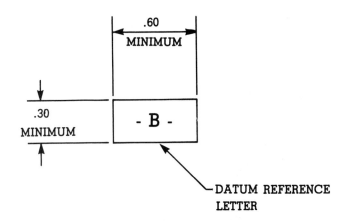

FIGURE 8.12
Datum identifying symbol.

8.8.3 The Basic Dimension Symbol

The symbol method of identifying a basic dimension is by enclosing it in a frame, as illustrated in Figure 8.13.

8.8.4 The Datum Target Symbol

The **datum target symbol** is a circle divided into four quadrants, as shown in Figure 8.14. The letter placed in the upper left quadrant identifies its associated datum feature, and the numeral placed in the lower right quadrant identifies the target.

8.9 COMBINED SYMBOLS

When necessary, individual symbols, datum reference letters, and the desired tolerances are systematically brought together to express tolerances by the use of **combined symbols**.

8.9.1 The Feature Control Symbol

A **positional tolerance** defines a zone within which the center, axis, or center plane of a feature of size is permitted to vary from true position. A **form tolerance** controls straightness, flatness, circularity, and cylindricity. Either tolerance may be stated by means

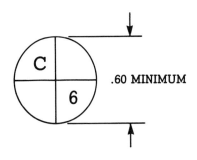

FIGURE 8.14
Datum target symbol.

of a **feature control symbol**, which consists of a frame containing the geometric symbol followed by the allowable tolerance, as illustrated in Figure 8.15(a). A vertical line separates the symbol from the tolerance.

Also, where applicable, the tolerance is preceded by the diameter symbol and followed by the symbol for MMC (maximum material condition) or RFS (regardless of feature size), as presented in Figure 8.15(b).

8.9.2 The Feature Control Symbol Incorporating Datum Reference

If a positional or form tolerance is related to a datum, the relationship is stated within the feature control symbol by placing the datum reference letter between the geometric characteristic symbol and the tolerance, as shown in Figure 8.16. Vertical lines separate these areas.

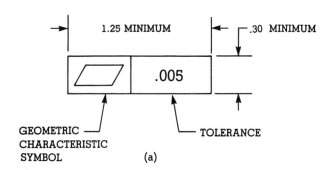

(a)

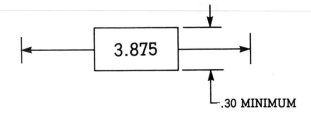

FIGURE 8.13
Basic dimension symbol.

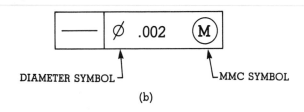

(b)

FIGURE 8.15
Feature control symbols.

FIGURE 8.16
Feature control symbols
incorporating datum
references.

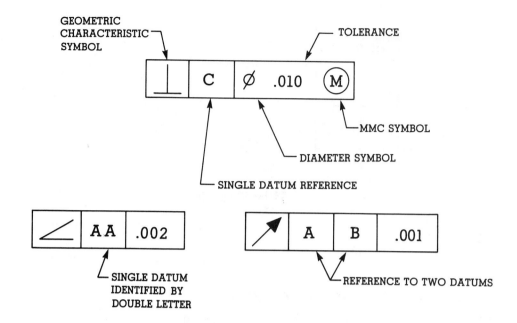

FIGURE 8.17
Order of precedence of datum
references.

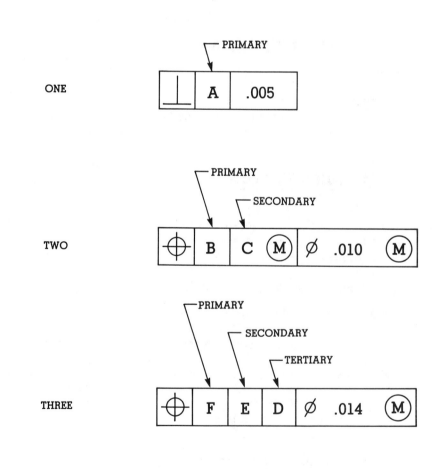

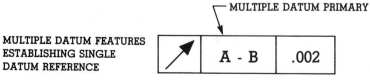

When required, the datum reference letter entry includes the symbol for MMC or RFS, and the length of the frame is increased as necessary to accommodate the additional data.

Each datum reference letter is entered in the desired order of precedence, from left to right, in the feature control symbol. Datum reference letter entries do not have to be in alphabetical order. Where a single datum reference is established by multiple datum features, the datum reference letters are separated by a dash, as shown in Figure 8.17.

A composite feature control symbol is used where more than one tolerance of a given geometric characteristic applies to the same feature. A single entry of the geometric characteristic symbol is followed by each tolerance requirement, one above the other and separated by a horizontal line, as depicted in Figure 8.18.

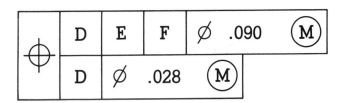

FIGURE 8.18
Composite feature control symbol.

8.9.3 The Combined Feature Control and Datum Identifying Symbol

When a feature is controlled by a positional or form tolerance and serves as a datum feature, the feature control and datum identifying symbols are combined, as indicated in Figure 8.19.

In such cases, the length of the frame for the datum identifying symbol may be either the same as that of the feature control or .60 inch minimum.

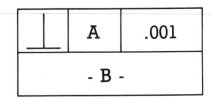

FIGURE 8.19
Combined feature control and datum identifying symbol.

8.9.4 The Combined Feature Control and Projected Tolerance Zone Symbol

Where a positional or perpendicularity tolerance is specified as a projected tolerance zone, a frame containing the projected height followed by the appropriate symbol is placed beneath the feature control symbol, as illustrated in Figure 8.20.

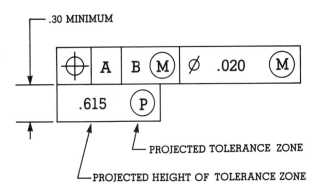

FIGURE 8.20
Combined feature control symbol with a projected tolerance zone.

8.10 THE FORM AND PROPORTION OF SYMBOLS

The construction, form, and proportion of individual geometric tolerancing symbols are represented in Figure 8.21.

8.11 THE PLACEMENT AND APPLICATION OF GEOMETRIC CHARACTERISTIC SYMBOLS

It is significant that the feature control symbol be associated with the feature(s) being toleranced by one of the following methods, which are illustrated in Figure 8.22.

1. Adding the symbol to a note or dimension pertaining to the feature.
2. Running a leader from the feature to the symbol.
3. Attaching a side, end, or corner of the control symbol frame to an extension line from the feature.
4. Attaching a side or end of the frame to the dimension line pertaining to the feature.

FIGURE 8.21
Form and proportion of symbols.

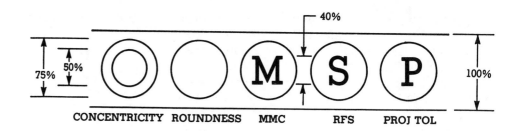

CONCENTRICITY ROUNDNESS MMC RFS PROJ TOL

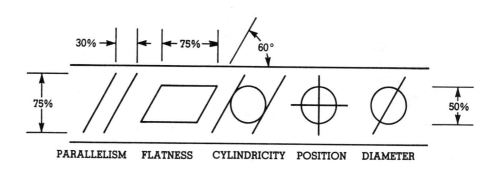

PARALLELISM FLATNESS CYLINDRICITY POSITION DIAMETER

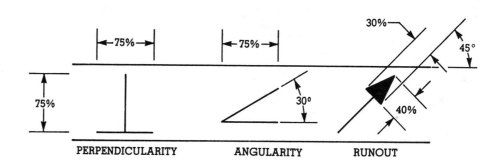

PERPENDICULARITY ANGULARITY RUNOUT

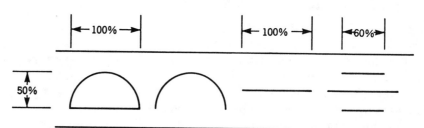

PROFILE - SURFACE PROFILE - LINE STRAIGHTNESS SYMMETRY

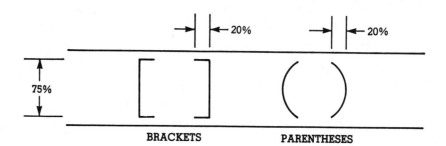

BRACKETS PARENTHESES

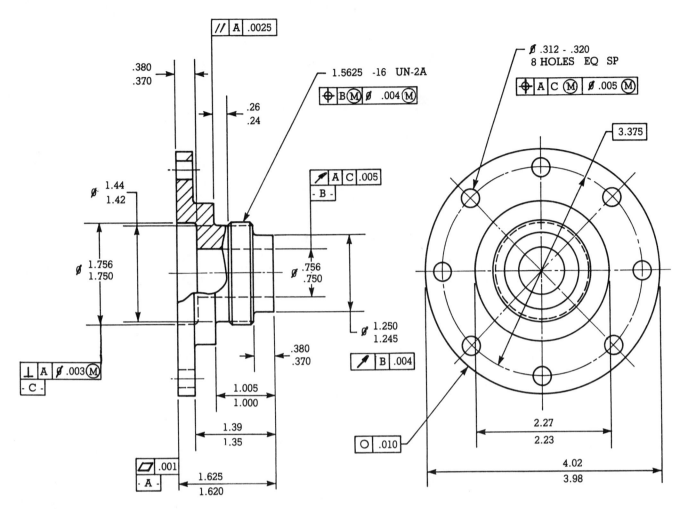

FIGURE 8.22
Application of geometric characteristic symbols.

8.12 SUMMARY

No dimension on a drawing can be held without some variance. This variance is called a tolerance. This section explained the need and purpose for tolerances. Examples of dimensions to be controlled were identified, and several technical terms related to the topic were defined. In addition, tolerancing methods, tolerance expression, interpretation of limits, and accumulation of tolerances were discussed.

An in-depth coverage of the usage of the American National Standards Institute (ANSI) specification Y14.5M-1982 was presented. Information on symbology characteristics for geometric tolerancing was emphasized, and examples were provided on the meaning, application, and placement of geometric tolerance characteristics on drawings. Practice exercises for this section are presented on pages 8A through 8G in the workbook.

8.13 **REVIEW EXERCISES**

1. Define "tolerance."

2. Why is tolerancing important on a drawing?

3. Define the tolerancing terms listed below:
Basic size:

Limit dimensions:

Allowance:

Fit:

Interference allowance:

Clearance allowance:

4. Explain the term "geometric tolerance."

5. What American National Standards Institute specification is used throughout industry for dimensioning and tolerancing?

6. Match the geometric characteristic symbols below with the correct terms:

a. _____   _____ perpendicularity

b. ⊚   _____ angularity

_____ roundness

c. ▱   _____ flatness

d. ∠   _____ straightness

_____ concentricity

e. ◯

f. ⊥

g. ⌀

h. ∥

7. What geometric symbols are used for the following terms?

Maximum material condition (MMC) _____

Regardless of feature size (RFS) _____

Diameter (DIA) _____

Basic (BSC) _____

8. Explain the difference between a bilateral and a unilateral tolerance. Give examples of each.

9. Compare tolerance accumulations resulting from chain dimensioning, datum dimensioning, and direct dimensioning.

10. Which three letters of the alphabet should not be used as datum reference letters? Why?

11. Define "feature control symbol."

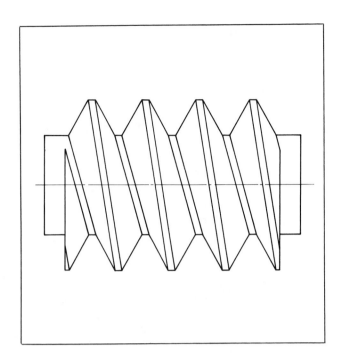

Section **9**

The Screw Thread and the Fastener

LEARNER OUTCOMES

The student will be able to:

- State the purpose of the screw thread
- Identify basic screw thread forms used in industry
- List and define several screw thread terms
- Represent and designate screw threads on technical drawings
- Determine screw lengths given engineering data
- Calculate clearance hole sizes for screws
- Demonstrate learning through the successful completion of review and practice exercises

9.1 THE PURPOSE OF THE SCREW THREAD

The predominant method for fastening together two or more parts which may require occasional disassembly and reassembly because of wear, for adjustment, or for the transmission of power, is the **screw thread**. Simply defined, the screw thread is a spiral-shape groove cut into the surface of a bar, rod, or cylinder. For proper operation there must be two mating, movable components: one with an internal thread and one with an external thread. The simplest types of mating parts having internal and external threads are a screw and a nut. Figure 9.1 illustrates a hexagonal head cap screw (external thread) and a hexagonal nut (internal thread), which are commonly used in industrial fastening applications.

The screw thread is utilized in the design, fabrication, production, and operation of most products. It is therefore necessary for the drafter to be able to recognize screw threads and to represent them graphically and specify them accurately on engineering drawings. There are several different types of thread forms, including the acme, American National, square, unified, buttress, knuckle, worm, and sharp V threads. These thread forms are depicted in Figure 9.2. Each serves a specific purpose.

1. **The acme screw thread** The acme screw thread is a thread form designed for high stress in traversing motion and for power transmission.
2. **The American National thread** The American National thread is suitable for use on bolts, screws, nuts, and for general use on polymers and soft metals. American National threads are designated as N, NC, NF, NEF, or NS.
3. **The square thread** The square thread is used primarily for the transmission of power. It is so designated because its teeth are square and at right angles to the axis. The square thread is an older thread form which has been replaced by the acme form.
4. **The unified thread** The unified thread is the most widely used of all thread forms and is usually referred to as the **Unified National** thread. It has been approved by standards organizations in Canada, the United States, and Great Britain. Its applications are the same as those of the American National thread, and it is mechanically interchangeable with American National threads of the same diameter and pitch. Unified threads are designated as UN, UNC, UNF, UNEF, UNS, or UNM.
5. **The buttress thread** The buttress thread is a form used in applications involving exceptionally high stress, in one direction only, along the thread axis.
6. **The knuckle thread** The knuckle thread is usually found on glass and plastic jars and on electric light bulbs and sockets. Threads for jar lids are usually produced from rolling sheet metal.
7. **The worm thread** The worm thread, like the acme and square threads, is used for power transmission.
8. **The sharp V thread** The sharp V thread has a sharp or pointed crest which makes it useful for applications that involve assembly and/or adjustment.

Because multinational interchangeability of parts is required for large companies, the newest thread standard being used throughout the world is

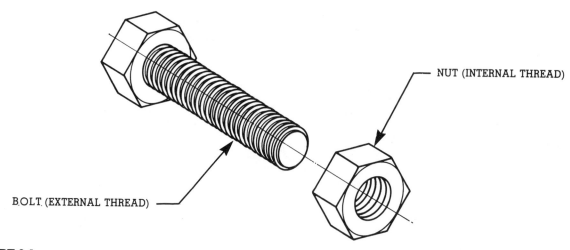

FIGURE 9.1
Examples of components having internal and external threads.

FIGURE 9.2
Screw thread forms.

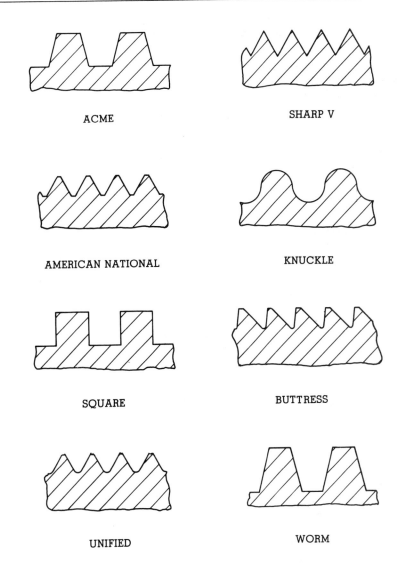

ACME

SHARP V

AMERICAN NATIONAL

KNUCKLE

SQUARE

BUTTRESS

UNIFIED

WORM

the **ISO metric thread form**. It has the same profile as that of the Unified National thread except that its thread depth is not as great, and it is used in the same applications. Figure 9.3 shows a comparison between the ISO metric and Unified National thread forms.

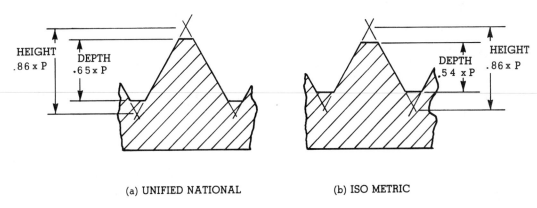

HEIGHT
.86 x P

DEPTH
.65 x P

DEPTH
.54 x P

HEIGHT
.86 x P

(a) UNIFIED NATIONAL

(b) ISO METRIC

FIGURE 9.3
Comparison of ISO metric and Unified National thread forms.

9.2 SCREW THREAD TERMINOLOGY

To be able to speak intelligently about screw threads and fastening hardware with engineering personnel and vendors, the drafter needs to be familiar with the basic but important terms listed below, some of which are illustrated in Figure 9.4.

1. **Crest** The surface that joins the sides of a thread and is the farthest from the cylinder or cone from which it projects.
2. **Depth of thread** The distance between the crest and the root of the thread, as measured normal to the axis.

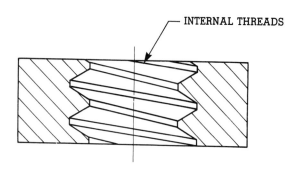

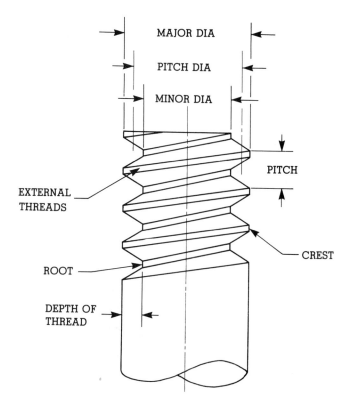

FIGURE 9.4
Thread terminology.

3. **External thread** A thread located on the external periphery of a rod, cylinder, or cone.
4. **Form of thread** That portion of a thread's profile (cross section) in an axial plane for a length of one pitch.
5. **Internal thread** A thread located on the internal surface of a cone or cylinder.
6. **Lead** The distance a threaded part moves axially, with respect to a fixed mating part, in one complete revolution.
7. **Left-hand thread** A thread that, when viewed axially, winds in a counterclockwise and receding direction. A left-hand thread is designated LH on a drawing.
8. **Major diameter** The largest or external diameter of a screw thread.
9. **Minor diameter** The smallest or root diameter of a screw thread.
10. **Pitch** The distance between corresponding points on adjacent thread forms measured parallel to the axis.
11. **Nominal thread size** The general designation given for the identification of threads.
12. **Pitch diameter** The equivalent of the diameter of an imaginary cylinder whose surface cuts the thread forms where the width of the thread and groove are equal.
13. **Root** The edge or surface that joins the sides of adjacent thread forms and coincides with the cylinder or cone from which the side projects.
14. **Right-hand thread** A thread that, when viewed axially, winds in a clockwise and receding direction. On drawings, threads are always considered to be right-hand threads unless otherwise noted.
15. **Thread class** The amount of tolerance or tightness of fit between internal and external threads (also referred to as **class of fit**). A designation of **fit** consists of a "1," a "2," or a "3" (tightest fit) followed by an "A" (external thread) or a "B" (internal thread). Refer to Appendix H for thread class information.
16. **Threads per inch (TPI)** The number of threads in one inch of length of screw threads. Measurement can be accomplished with a scale or with a thread gage.
17. **Thread series** For the Unified National thread, the classification of a thread which indicates the number of threads per inch for a specified diameter. For example, the words "coarse," "fine," and "extra fine" and the numbers 8, 12, and 16 are used to describe the thread series. Appendix H provides information on standard thread series.

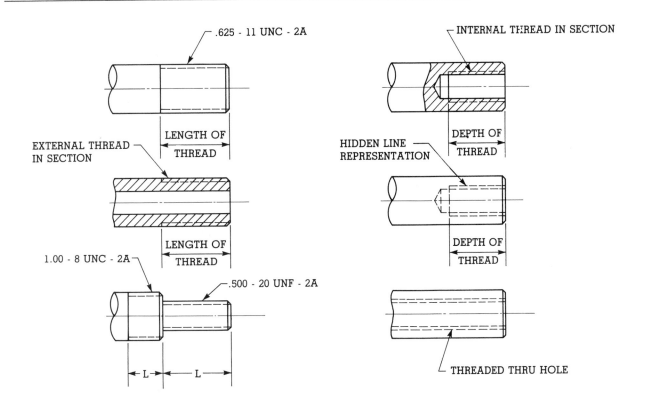

FIGURE 9.5
Simplified representation of screw threads.

9.3 DRAWING APPLICATION OF SCREW THREADS

In the application of screw threads to a drawing, there are two areas to consider: (1) **thread representation** (how threads are shown graphically on the drawing); and (2) **thread designation** (identification of threads by means of text information, according to some accepted standard).

9.3.1 Thread Representation

There are three types of representations used by industry to portray all screw thread forms on technical drawings: the **detailed representation**, the **schematic representation**, and the **simplified representation**. Of the three, only the simplified representation will be covered here, because it is the most commonly used method and is preferred by the Department of Defense specification MIL-STD-9 and the American National Standards Institute specification ANSI Y14.6. Figure 9.5 illustrates how the simplified representation is drawn.

9.3.2 Thread Designation

Thread designation is text information provided on a drawing for identification of standard series threads and consists of three groups of alphanumeric characters, separated by dashes, indicating nominal size; number of threads per inch and thread series symbol; and thread class. Figure 9.6 shows a typical thread designation in the decimal-inch system.

The **nominal size** of threads is designated in decimal form. No tolerance is noted, and no toler-

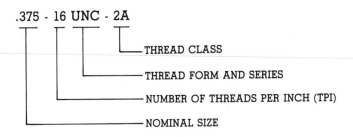

FIGURE 9.6
Basic thread designation.

.250 - 20 UNC - 3A   (EXTERNAL THREAD)

.250 - 20 UNC - 3B   (INTERNAL THREAD)

FIGURE 9.7
Nominal thread size.

.312 - 18 UNC - 2A

.312 - 18 UNC - 2B

FIGURE 9.8
Designation for the number of threads per inch.

ance is applicable in the example shown in Figure 9.7, where .250 is the nominal size.

The **number of threads** indicates the number of **threads per inch** (TPI). In Figure 9.8, the "18" denotes the number of threads per inch.

There are four **unified thread series** which have been standardized. Table 9.1 identifies thread forms and series for the Unified National thread. Each form has somewhat specific applications for use in industry.

Thread classes differ from each other by the amount of tolerance, or by the amount of tolerance and allowance. For those threads which are categorized under the unified form, classes 1A, 2A, and 3A apply only to external threads, and classes 1B, 2B, and 3B apply only to internal threads. For the American National thread form, the class of thread for both internal and external threads is indicated by a number only. Figure 9.9 shows how thread classes are designated for both the unified and American National forms.

Hand designation indicates whether screw threads are right hand or left hand. Threads are right

.500 - 20 UNF - 2A   (EXTERNAL THREAD)

.500 - 20 UNF - 2B   (INTERNAL THREAD)

(a) UNIFIED FORM

.500 - 20 UNF - 2   (INTERNAL OR EXTERNAL THREAD)

(b) AMERICAN NATIONAL FORM

FIGURE 9.9
Designation of thread class for the Unified and American National thread forms.

hand unless "LH" (for left hand) is included in the callout (see Figure 9.10).

9.3.3 Metric Thread Designation

Basic metric thread designation differs from the decimal-inch method. ISO metric threads are designated by the letter "M" followed by the nominal size (in millimeters), the sign "X," the pitch (in millimeters), and the tolerance class. Figure 9.11 shows a typical ISO metric thread designation.

The coarse pitch ISO metric threads may be designated by only the letter "M" and the nominal size in millimeters. No tolerance is given at this point. An appropriate example is shown in Figure 9.12.

.250 - 32 UNEF - 2A - LH

FIGURE 9.10
Left hand thread designation.

TABLE 9.1
Unified thread series usage.

SERIES	USAGE
UNC (UNIFIED NATIONAL COURSE)	SCREWS, NUTS, BOLTS, THREADS IN SOFT METAL AND POLYMERS FOR GENERAL USE.
UNF (UNIFIED NATIONAL FINE)	SCREWS, NUTS, BOLTS, THAT REQUIRE A BETTER FIT; FOR USE ON AUTOMOTIVE PARTS, ALSO, WHERE ADJUSTMENT IS REQUIRED
UNEF (UNIFIED NATIONAL EXTRA FINE)	FOR TUBING, THIN NUTS, AND APPLICATIONS IN WHICH THE LENGTH OF THREAD IS RELATIVELY SHORT
UN (CONSTANT PITCH)	FOR SPECIAL PURPOSE APPLICATIONS FOR WHICH LARGE DIAMETERS AND OTHER THREADS DO NOT APPLY

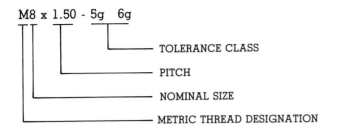

FIGURE 9.11
Metric thread designation.

M16

FIGURE 9.12
Course pitch metric designation.

The tolerance system for ISO metric screw threads provides for allowances and tolerances indicated by a **tolerance class**. There are three tolerance grades—grade 4, grade 6, and grade 8—all of which reflect the size of the tolerance, categorized as follows:

- Grade 8 tolerance is closest to the unified class 1A and class 1B fits.
- Grade 6 tolerance is closest to the unified class 2A and class 2B fits.
- Grade 4 tolerance is closest to the unified class 3A and class 3B fits.

ISO has also established the amount of tolerance by providing a series of tolerance position symbols for external and internal threads, as follows:

- For external threads (screws, bolts), "e" indicates large allowance, "g" indicates small allowance, and "h" indicates no allowance.
- For internal threads (nuts), "G" indicates small allowance and "H" indicates no allowance.

The symbols above are used after the tolerance grade. For example, "6g" indicates a "medium" tolerance grade with "small" allowance for an external thread.

The ISO tolerance class of fit is determined by selecting one of three qualities (fine, medium, or coarse) combined with one of three lengths of engagement (short, S; normal, N; or long, L). The use of tolerance positions "g" for external threads and "H" for internal threads are preferred.

Complete external and internal designations for ISO metric screw threads are illustrated in Figure 9.13.

9.3.4 Dimensioning of Screw Threads

Dimensioning and tolerancing of screw threads should generally follow the practices presented ear-

lier in this section. However, there are some special conditions and applications that need be considered only in dimensioning of screw threads. They include:

1. In those cases where threads are shown as a portion of a greatly enlarged detail, the precise representation of thread geometry might be considered, as depicted in Figure 9.14.
2. In dimensioning of thread length, the dimension should include the length of threads having full form, as illustrated in Figure 9.15. Incomplete threads shall be outside or beyond the length specified. Should there be a reason to limit the number of incomplete threads, the overall thread length may be added in addition to the full thread length dimension, as represented in Figure 9.16.
3. If a definite length of unthreaded section is of importance, equally or more so than the threaded length, it is normally dimensioned as indicated in Figure 9.17.
4. Dimensioning of an undercut or tool relief is performed as shown in Figure 9.18.
5. When the wall thickness at drill point for blind tapped holes is critical, the tap drill point depth or wall thickness should be dimensioned as illustrated in Figure 9.19.
6. The screw thread designation is added with a leader line, as depicted in Figure 9.20.

9.4 THE FASTENER

Fasteners, both threaded and unthreaded, are an important, fundamental method for joining parts together, which makes them an essential component of almost every design. There are three basic categories of fasteners: the **removable type**, the **semipermanent type**, and the **permanent type**. Removable fasteners are classified as those that can be removed easily with hand tools without damaging mating parts. Typical examples include the many types of bolts and nuts used around the home and in industry. Semipermanent fasteners can also be removed easily, but this may result in some damage to the fastener. Examples of this type of fastener include cotter pins, taper pins, grooved pins, dowel pins, and special self-locking nuts with nylon inserts. Permanent fasteners are those that are put in place to remain permanently installed and are not removed or disassembled for routine or periodic maintenance work. A rivet is an example of a permanent fastener.

9.4.1 The Threaded Fastener

Threaded or screw-type fasteners are the most commonly utilized fastening devices. Figure 9.21

FIGURE 9.13
ISO metric screw thread designation for external and internal threads.

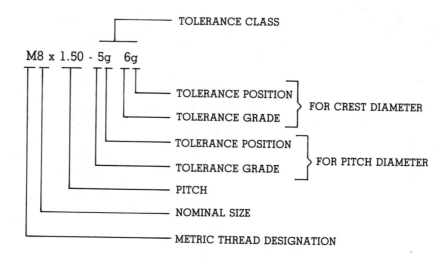

(a) EXTERNAL THREAD

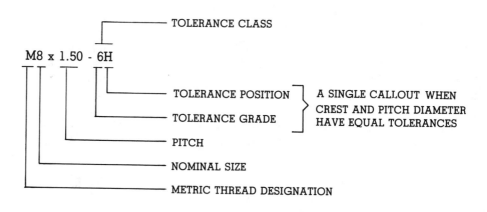

(b) INTERNAL THREAD

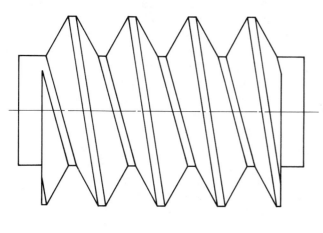

SCALE 4:1

FIGURE 9.14
Thread representation for large detail.

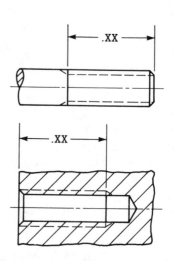

FIGURE 9.15
Thread length dimensioning.

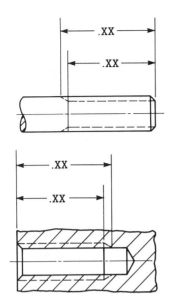

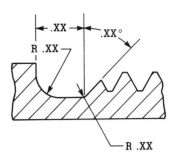

FIGURE 9.18
Dimensioning of an undercut.

FIGURE 9.16
Overall thread length dimensioning.

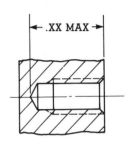

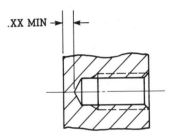

FIGURE 9.19
Dimensioning for critical depth.

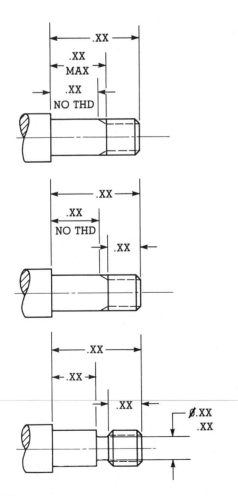

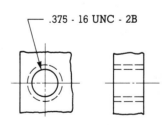

.375 - 16 UNC - 2B

INTERNAL

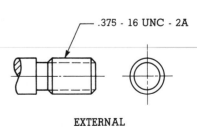

.375 - 16 UNC - 2A

EXTERNAL

FIGURE 9.17
Dimensioning of threaded and unthreaded lengths.

FIGURE 9.20
Screw thread designation.

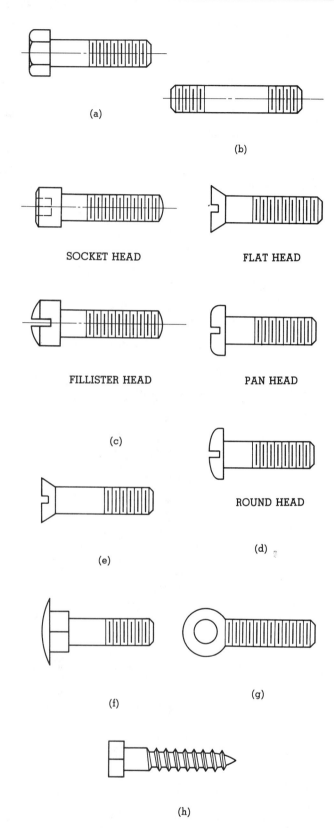

FIGURE 9.21
(a) Through bolt; (b) Stud; (c) Cap screws; (d) Machine screws; (e) Stove bolt; (f) Carriage bolt; (g) Eyebolt; (h) Lag screw.

provides examples of threaded fasteners used in industry as well as in the building trades.

The **through bolt** shown in Figure 9.21(a) usually has a square or hex-shape head at one end and threads at the other and is often used to join two parts together by being passed through each of them and secured with a nut. A washer is sometimes required at the head end or the nut end, and a wrench is used to tighten or loosen the nut or bolt.

The **stud** is required when space is needed between two mating parts or when a through bolt is not suitable for parts that must be removed often. Note in Figure 9.21(b) that the stud is threaded on both ends. It may be of a different thread series for the required application.

The **cap screw**, which is illustrated in Figure 9.21(c), is most used in applications that call for it to be passed through a clearance hole in the nearest part and screwed into a threaded hole in the other part. Cap screws are usually finished to improve their appearance. Popular head shapes include the hex head, the fillister head, and the socket head.

The **machine screw** is a small threaded fastener used in light-to-medium-weight applications. It functions with or without a nut, as does the cap screw. Several types of heads are available, as depicted in Figure 9.21(d).

The **stove bolt** is an inexpensive fastener that is not made to close tolerances. A loose fit between the bolt and nut makes assembly easy. An example is shown in Figure 9.21(e).

The **carriage bolt**, shown in Figure 9.21(f), is characterized by a square section immediately under its head. The head is plain—that is, it doesn't have a slot or groove to accommodate a tool. A square nut is often used at the threaded end. As the nut is tightened, the carriage bolt is held stationary by the square section. This section seats itself in a square cutout when used in metal and wedges into the wood when used for fastening wood to metal or wood to wood.

The **eyebolt** is used for lifting purposes—specifically, for hoisting pieces of equipment. It is often used for lifting transportable, mobile, or air-dropped military equipment. The eyebolt is often attached with a through hole and nut or screwed into a threaded hole. Figure 9.21(g) illustrates the eyebolt.

The **lag screw**, shown in Figure 9.21(h), has several uses, but a major industrial application is the fastening of machinery to pallets or bases prior to shipment. A lag screw resembles a bolt at the head end and a wood screw at the threaded end. Heads are either hexagonal or square in shape and, therefore, require a wrench for tightening.

9.4.1.1 Screw Length Selection

It is the drafter's responsibility to determine the lengths of hardware. Given the combination of material thickness, washers, and nuts, there are some basic rules to be followed in determining the correct screw length for a given application.

Grip length is the sum of any combination of material thickness plus any hardware under the screw head, as shown in Figure 9.22. Do not add the thickness of any hardware, or the equivalent of 1½ threads below the material, because this has been allowed for in the screw length chart presented in Table 9.2. To determine proper screw length, calculate the grip as illustrated in Figure 9.23.

If the calculated grip is not shown in Table 9.2 under the appropriate screw size, use the next larger grip value. The proper screw length is found on the same line in the column to the extreme left, "Screw Length." *Example:* Since .215 is not listed under the #8-32 screw size in Table 9.2, use the next larger grip, which is .236. Note that this corresponds to a screw length of .500 inch, which is the correct screw length for this particular application. When a washer is deleted, as represented in Figure 9.24, subtract its thickness (.049) from the grip value (.215) to obtain a correct grip of .166 inch. Since a grip of .166 is not listed in Table 9.2 under the 8-32 screw size, use the

TABLE 9.2
Screw length chart.

SCREW SIZE	2-56	4-40	6-32	8-32	10-32	1/4	5/16	3/8
SCREW LENGTH								
.125	—	—	—	—	—	—	—	—
.188	.056	—	—	—	—	—	—	—
.250	.118	.068	.011	—	—	—	—	—
.312	.180	.130	.073	.048	.041	—	—	—
.375	.243	.193	.136	.111	.104	—	—	—
.438	.306	.256	.199	.174	.167	.012	—	—
.500	.368	.318	.261	.236	.229	.074	—	—
.625	.493	.443	.386	.361	.354	.199	.128	.039
.750	.618	.568	.511	.486	.479	.324	.253	.164
.875	.743	.693	.636	.611	.604	.449	.378	.289
1.000	—	.818	.761	.736	.729	.574	.503	.414
1.250	—	1.068	1.011	.986	.979	.824	.753	.664
1.500	—	1.318	1.261	1.236	1.229	1.074	1.003	.914
1.750	—	—	1.511	1.486	1.479	1.324	1.253	1.164
2.000	—	—	1.761	1.736	1.729	1.574	1.503	1.414
2.250	—	—	—	1.986	1.979	1.824	1.753	1.664
2.500	—	—	—	2.236	2.229	2.074	2.003	1.914
2.750	—	—	—	2.486	2.479	2.324	2.253	2.164
3.000	—	—	—	2.736	2.729	2.574	2.503	2.414

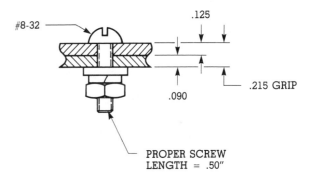

FIGURE 9.23
Grip length calculation.

next larger grip (.174), which specifies a screw length of .438 inch for this application.

9.4.1.2 Selection of Clearance Hole Diameter

When a **clearance hole** is required for a single screw, Appendix C may be used for reference. However, if two or more fasteners are needed to join mating parts, the clearance holes and tolerances will have to be calculated to ensure that an interference fit does not result. When this situation occurs, the following practice may be followed. Four conditions will be described and examples of each presented.

The examples provided are for a #8 pan head screw (.164-inch diameter) with a tolerance of +.010, −.010 inch (T = .020). The terms in the solutions are as follows:

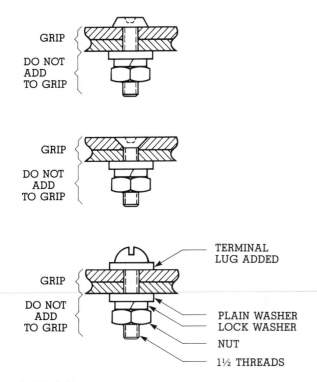

FIGURE 9.22
Examples of grip length.

FIGURE 9.24
Corrected grip length.

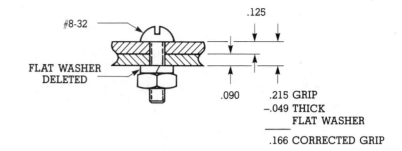

.215 GRIP
−.049 THICK
FLAT WASHER
.166 CORRECTED GRIP

M = maximum screw diameter
T = sum of **bilateral** tolerance between holes
D = minimum diameter of clearance hole

$$M + \frac{T}{2} = D$$

Therefore, $.164 + \frac{.020}{2} = .174$ diameter

Table 9.3 should be used for the problem solution. Note that it is arranged to show the nearest drill size for conditions No. 1 through No. 4.

Condition No. 1: Pattern of two holes with clearance holes in both mating parts, as illustrated in Figure 9.25.

The closest hole diameter for a No. 8 screw in Table 9.3 indicates that each hole should be .177 inch in diameter (No. 16 drill).

Condition No. 2: Pattern of two holes with a clearance hole in one part and a tapped hole or

TABLE 9.3
Clearance holes for screws.

SIZE OF SCREW	CONDI- TION	±.005 DRILL SIZE	±.010 DRILL SIZE	±.015 DRILL SIZE	±.020 DRILL SIZE	±.025 DRILL SIZE	±.030 DRILL SIZE
NO. 2 (.086 DIA.)	1	.0935(42)	.096(41)	.1015(38)	.1065(36)	.111(34)	.116(32)
	2	.096(41)	.1065(36)	.116(32)	.1285(30)*,#	.136(29)*,#	.147(26)*,#
	3	.1015(38)	.116(32)	.1285(30)*,#	.144(27)*,#	.157(22)*,#	.173(17)*,#
	4	.116(32)	.144(27)*,#	.173(17)*,#	.199(8)*,#	.228(1)*,#	.257(F)*,#
NO. 4 (.112 DIA.)	1	.120(31)	.125(1/8)	.1285(30)	.136(29)	.1405(28)	.144(27)
	2	.125(1/8)	.136(29)	.144(27)	.152(24)*	.166(19)*	.173(17)*,#
	3	.1285(30)	.1405(28)	.154(23)*	.1695(18)*,#	.182(14)*,#	.196(9)*,#
	4	.1405(28)	.1695(18)*,#	.196(9)*,#	.228(1)*,#	.257(F)*,#	.2811(K)*,#
NO. 6 (.138 DIA.)	1	.144(27)	.1495(25)	.154(23)	.159(21)	.166(19)	.1695(18)
	2	.1459(25)	.159(21)	.1695(18)	.180(15)	.189(12)*	.199(8)*
	3	.152(24)	.166(19)	.180(15)	.196(9)*	.209(4)*,#	.228(1)*,#
	4	.166(19)	.196(9)*	.228(1)*,#	.250(1/4)*,#	.2811(K)*,#	.3125(5/16)*,#
NO. 8 (.164 DIA.)	1	.1695(18)	.177(16)	.180(15)	.185(13)	.189(12)	.196(9)
	2	.177(16)	.185(13)	.196(9)	.204(6)	.2187(7/32)	.228(1)*,#
	3	.180(15)	.1935(10)	.209(4)	.221(2)*	.2344(15/64)*	.250(1/4)*,#
	4	.1935(10)	.221(2)*	.250(1/4)*,#	.277(J)*,#	.3125(5/16)*,#	.332(Q)*,#
NO. 10 (.190 DIA.)	1	.196(9)	.201(7)	.2055(5)	.213(3)	.2187(7/32)	.221(2)
	2	.201(7)	.213(3)	.221(2)	.234(A)	.242(C)	.250(1/4)
	3	.204(6)	.2187(7/32)	.234(A)	.246(D)	.261(G)	.277(J)*
	4	.2188(7/32)	.246(D)	.277(J)*	.302(N)*,#	.332(Q)*,#	.3594(23/64)*,#
1/4" (.250 DIA.)	1	.257(F)	.261(G)	.266(H)	.272(I)	.277(J)	.2811(K)
	2	.261(G)	.272(I)	.2811(K)	.290(L)	.302(N)	.3125(5/16)
	3	.2656(17/64)	.2811(K)	.295(M)	.3125(5/16)	.323(P)	.339(R)
	4	.2811(K)	.3125(5/16)	.339(R)	.368(U)*	.3906(25/64)*	.4219(27/64)*,#

NOTE - DRILL SIZE IS CLOSEST SIZE WITHOUT TOLERANCE
CONSIDERATION
* MEDIUM LOCKWASHER WILL DROP INTO CLEARANCE HOLE
UNDER WORST CONDITION
HEAD OF SCREW WILL NOT COVER HOLE UNDER WORST CONDITION

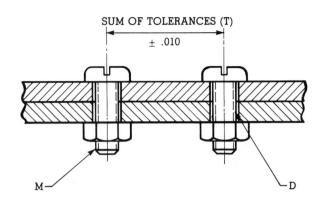

FIGURE 9.25
Selection of clearance hole diameter (condition No. 1).

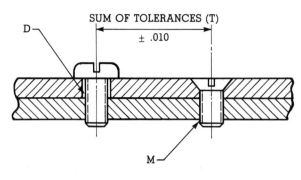

FIGURE 9.26
Selection of clearance hole diameter (condition No. 2).

countersunk hole in the mating part, as portrayed in Figure 9.26.

$$M + T + D$$
Therefore, $.164 + .020 = .184$ diameter

The closest hole diameter for a No. 8 screw in Table 9.3 shows that each hole should be .185 inch in diameter (No. 13 drill).

Condition No. 3: Pattern of more than two holes with clearance holes in both parts, as depicted in Figure 9.27.

$$M + 1.4T = D$$
Therefore, $.164 + 1.4(.020) = .192$ diameter

The closest hole diameter in Table 9.3 is .1935 inch (No. 10 drill). Therefore, each hole should be .1935

inch in diameter to ensure that the parts mate properly.

Condition No. 4: Pattern of more than two holes with clearance holes in one part and tapped holes or countersunk holes in the mating part, as illustrated in Figure 9.28.

$$M + 2.8T = D$$
Therefore, $.164 + 2.8(.020) = .220$ diameter

The closest hole diameter for this application, according to Table 9.3, is .221 inch (No. 2 drill).

9.4.2 The Setscrew

The **setscrew** is another form of threaded fastener. It is used to prevent relative rotary motion between two parts, such as a shaft and a pulley. Setscrews can be removed easily and are classified according to the

FIGURE 9.27
Selection of clearance hole diameter (condition No. 3).

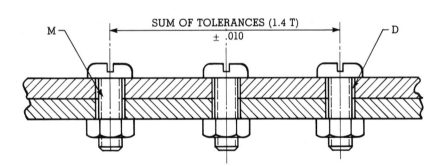

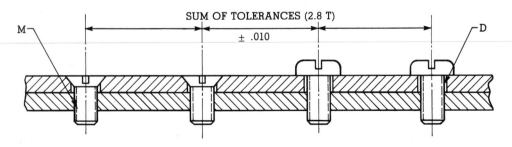

FIGURE 9.28
Selection of clearance hole diameter (condition No. 4).

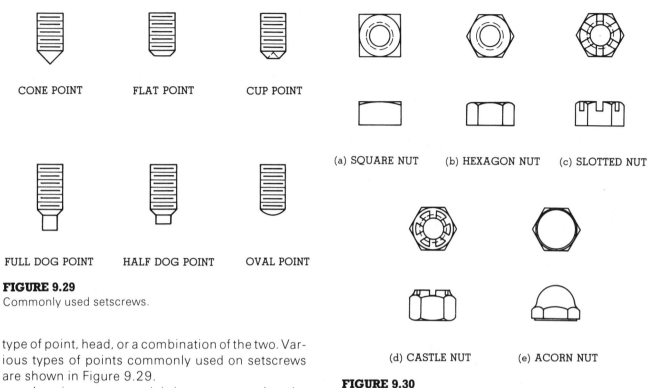

CONE POINT   FLAT POINT   CUP POINT

FULL DOG POINT   HALF DOG POINT   OVAL POINT

FIGURE 9.29
Commonly used setscrews.

(a) SQUARE NUT   (b) HEXAGON NUT   (c) SLOTTED NUT

(d) CASTLE NUT   (e) ACORN NUT

FIGURE 9.30
Nuts.

type of point, head, or a combination of the two. Various types of points commonly used on setscrews are shown in Figure 9.29.

In using setscrews, it is important to select the proper point for a specific application. The **cone point** is used in applications that require two parts to be joined in a permanent position relative to each other. The **flat point** is used when frequent adjustment is necessary between the two parts being secured. The **cup point** is the most widely used type. It allows rapid assembly with no preparation. The **full dog** and **half dog points** are used where members are to be joined permanently at a given location. For either type, a hole must be drilled to receive the point. The **oval point** is used in applications similar to those of the cup point.

9.4.3 The Nut

The **nut** is a fastener with internal threads that is used in conjunction with a screw or bolt for attaching parts together. Nuts are available in two wrench-type styles: hexagonal and square. Various types are utilized in industry depending on the specific application. Several different styles are represented in Figure 9.30.

9.4.4 The Rivet

The **rivet** is a permanent fastener. It is frequently used for fastening steel or aluminum sheet in boat and aircraft construction. A rivet consists of a short length of rod with some style of head at one end. Several head styles are available, as illustrated in Figure 9.31. Typical rivet materials are wrought iron, mild steel, aluminum alloys, and copper alloys. It is

important to emphasize that rivets cannot be removed without destroying the rivet and possibly damaging the mating parts.

9.4.5 The Washer

Washers are components used in conjunction with threaded fasteners. Plain **flat washers** are used to distribute the fastening load over an area larger than that of the head of a screw or a nut. A typical application is illustrated in Figure 9.32.

Spring lock washers are used to place an axial load between a bolt and nut so that vibration will not easily unthread the fastening.

9.4.6 The Retaining Ring

The **retaining ring** is a precision-type fastener that is used to secure components to shafts or to limit the movement of parts in an assembly. Applications vary from miniature assemblies to large construction equipment. Retaining rings are used in military hardware for space exploration as well as in business machines, automobiles, appliances, and other consumer products. Retaining rings are produced in two basic types: **external** and **internal**. Figure 9.33 shows the two basic types.

The external retaining ring for axial assembly is expanded over a shaft, stud, or similar part with a plier-type tool. The internal type for axial assembly is

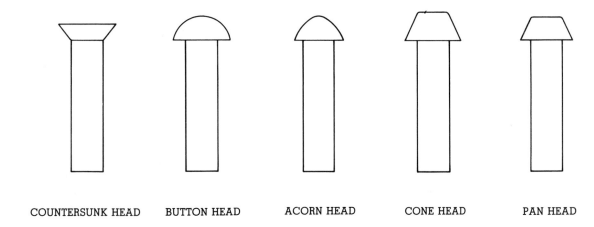

COUNTERSUNK HEAD   BUTTON HEAD   ACORN HEAD   CONE HEAD   PAN HEAD

FIGURE 9.31
Rivets.

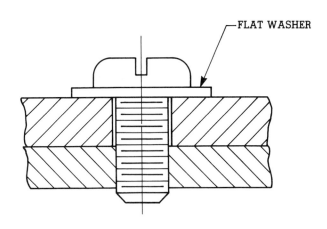

FIGURE 9.32
Application of a flat washer.

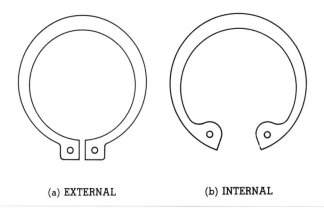

(a) EXTERNAL   (b) INTERNAL

FIGURE 9.33
Basic external and internal retaining rings.

compressed for insertion into a bore or a housing. Retaining rings are made of materials having excellent spring properties, such as carbon steel, stainless steel, and beryllium copper. Retaining rings are often used as an alternative to screws, cotter pins,

FIGURE 9.34
Examples of retaining rings. (Courtesy of Rotor Clip Company, Inc.)

rivets, collars, and machined shoulders. Annular grooves of the appropriate width and depth must be provided to accommodate the rings. Figure 9.34 displays examples of retaining rings.

9.4.7 Other Fastening Methods

Several additional practices are used for fastening mechanical components and parts together, including welding (both spot and heliarc), brazing, soldering, staking, and bonding with structural adhesives. Bonding with adhesives is becoming more and more popular as a fastening method. Epoxies have been used for many years for various industrial applications. Initially, glue and cement were applied where little strength was required. Today, because of great advances in chemistry, adhesives are frequently replacing riveting, brazing, soldering, and welding as the primary fastening method. Bonded brake linings are common on automobiles, and use of bonded aircraft parts and subassemblies is an accepted practice.

9.5 SUMMARY

Screw thread systems, including terminology, designations, internal and external threads, and drawing application, were covered in this section. Methods for determining screw length selection and clearance hole selection were outlined. Several examples of threaded fasteners, including through bolts, studs, cap and machine screws, stove bolts, carriage bolts, eyebolts, and lag screws, were provided, along with their definitions and applications. Washers and internal and external retaining rings were discussed. Finally, other fastening methods and practices were discussed. Emphasis was placed on structural adhesive bonding as a viable alternative to more traditional fastening methods. Practice exercises for this section are presented on pages 9A through 9G in the workbook.

9.6 REVIEW EXERCISES

1. What is a screw thread? How is it used?

2. List six types of screw threads and their general applications.

3. Insert the proper screw thread terms in the illustration below.

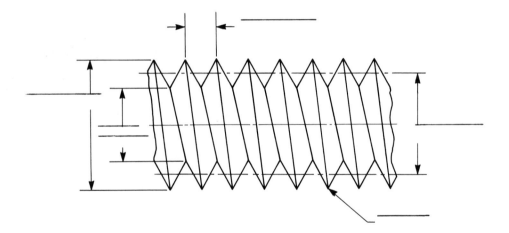

4. What is meant by "class of fit" for screw threads?

5. Explain the differences between thread representation and thread designation.

6. What do the following thread designations signify?
   a. .250 — 28 UNF — 2B

   b. .500 — 13 UNC — 3A

   c. M8 × 1.25 — 5g6g

7. Calculate the screw length for the following condition.

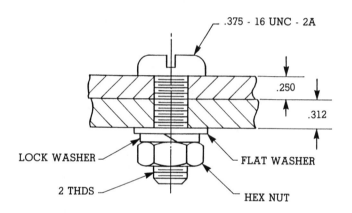

8. Determine the screw length for the condition illustrated below.

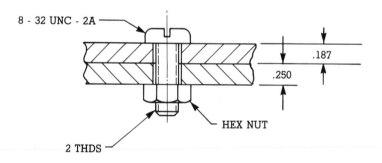

9. Determine the clearance hole sizes for the following conditions.
   a.

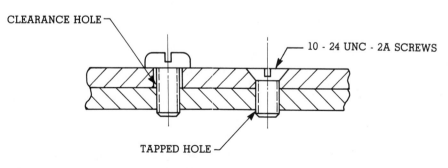

BILATERAL TOLERANCE = + .005, − .005

b.

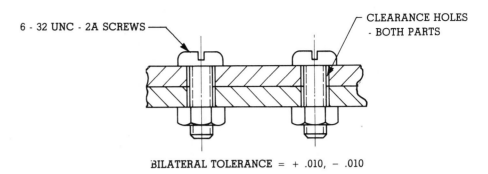

6 - 32 UNC - 2A SCREWS

CLEARANCE HOLES
- BOTH PARTS

BILATERAL TOLERANCE = + .010, − .010

10. What are the three basic categories of fasteners?

11. List ten types of fasteners.

12. Identify six types of threaded fasteners.

13. What is the purpose of the setscrew? Identify six types of points used on setscrews.

14. Name the wrench-type styles for nuts.

15. Where are rivets used?

16. Name two types of washers and their applications.

17. Define the term "retaining ring." Identify two types.

18. List five methods of fastening mechanical components and parts together.

Section 10

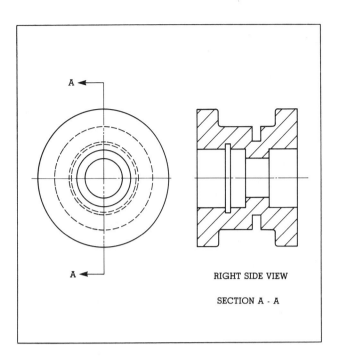

RIGHT SIDE VIEW

SECTION A - A

The Sectional View

LEARNER OUTCOMES

The student will be able to:

- State the purpose of the sectional view

- Identify the different types of sectional views

- Define the cutting plane line and demonstrate its use

- Draw several different sectional views given partial or incomplete information

- Demonstrate correct section lining procedures for single objects as well as those in assembly

- Explain why unconventional technical drafting practices are needed for drawing sections of webs, spokes, and ribs

- List eight basic guidelines for producing the sectional view

- Demonstrate learning through successful completion of review and practice exercises

10.1 PURPOSE

For an object with a complex internal shape and an interior construction that cannot be easily identified by an exterior view, a **sectional view** may be required. The purpose of the sectional view is to clearly show the internal form and detail of an object. A section usually eliminates the need for hidden lines while retaining the important outlines of the object. The sectional view is obtained by passing an imaginary line through some specific part of the object and presenting the object as though the part cut off by this line had been removed. An example is shown in Figure 10.1.

Figure 10.1 illustrates the front view of an object. To complete the drawing, the right-side view must also be shown so that the internal detail form of the object may be identified as portrayed in Figure 10.2. In the example shown there, one-fourth of the object has been removed. (This is known as a half section.)

Whenever practical, the sectional view should appear in correct orthographic projection with other views—that is, shown in the same relationship as the views identified in Figure 6.3.

Experienced drafters usually refer to the sectional view as a **section** or a **cross section**. Note also in Figure 10.2 the letter **A**. Upper-case letters such as **A, B, C, D**, and so on, are used to identify the section. The number of different letters used will depend on the complexity of the drawing and the number of sections required. MIL-STD-100, a U.S. Government military specification, states that letters should read horizontally, should not be underlined, and should be located in front of arrowheads. The section is also identified in the view which shows the cut-out portion of the object, as in Figure 10.2.

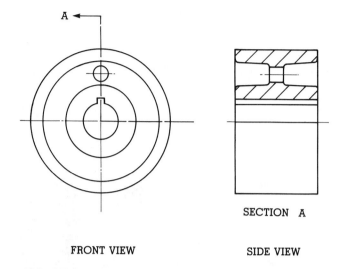

FRONT VIEW   SIDE VIEW

FIGURE 10.2
Sectional view.

10.2 THE CUTTING PLANE LINE

The cutting plane is determined by a **cutting plane line** which is used to indicate the location or position of the cutting plane for sectional views and the viewing position for partial views. This line is represented by a dashed or broken line in one of the three forms identified in Figure 10.3. Each is acceptable for use on sectional views whose line weight is thick, as shown in Figure 2.23. When used, each form should

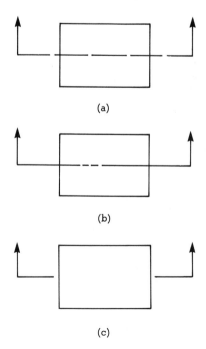

(a)

(b)

(c)

FIGURE 10.3
Cutting plane lines.

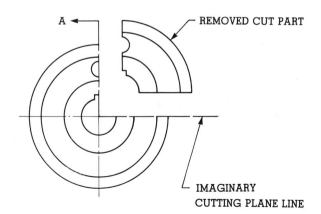

FIGURE 10.1
Imaginary cutting plane.

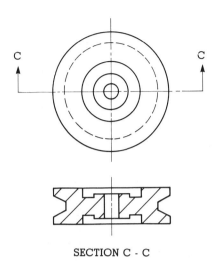

SECTION C - C

FIGURE 10.4
Omitting the cutting plane line.

be drawn so that it stands out clearly and cannot be confused with other lines.

Each cutting plane line shown in Figure 10.3 meets the approval of professional organizations for use on drawings. In Figure 10.3(a), the cutting plane line is formed by evenly spaced dashed lines which are turned 90 degrees at each end and terminated with arrowheads. The arrowheads indicate the viewing direction for the observer. Figure 10.3(b) shows the line as alternating long dashed lines and pairs of short dashed lines. The length of the long dashed lines may vary depending on the size and scale of the drawing. The line in Figure 10.3(c) shows the cutting plane but does not cross or make contact with the object. The cutting plane line may be omitted when it corresponds with the center line of an object and if it is obvious where the cutting plane lies, as depicted in Figure 10.4.

10.3 CONVENTIONS FOR SECTIONING

When a sectional view is created it exposes internal detail or a cut surface whose material should be identified and specified. This is accomplished through a conventional practice called **section lining**. There are several standard symbols used in lining a section which have been approved for use by the American National Standards Institute specification ANSI Y14.2. They are identified in Figure 10.5.

Section lining is also referred to as **crosshatching**. It (1) shows which surface has been theoretically cut, (2) illustrates it clearly, and (3) assists the observer in better understanding the internal shape

of the object. It may also show from which material the object is made if the symbols shown in Figure 10.5 are utilized. The symbols represented in Figure 10.5 are used for special applications with the exception of the one shown in Figure 10.5(a), which is the symbol for cast or malleable iron but also is a general-purpose symbol. Figure 10.6 is an enlarged view of the general-purpose section lining symbol, which is recommended for use in all review exercises and practice exercises for which section lining is required.

Section lining consists of sharp, uniformly spaced lines usually drawn at 45 degrees to the principal lines of the view. These lines are drawn parallel to each other and are rarely measured for spacing, but rather spaced by eye. The spacing of lines may vary from 1/16 to 1/4 inch depending on the size of the object and the scale to which the drawing is produced. Lines must be of uniform thickness and evenly spaced. Examples of acceptable and unacceptable section lining practices are presented in Figure 10.7. Refer to Figure 2.23 in Section 2 for proper weight of section lining.

When section lining is required for two or more mating parts, such as on an assembly drawing, the 45-degree section lines should be drawn in opposite directions, as illustrated in Figure 10.8.

If three or more adjacent parts are shown in section, the section lining should be drawn at 30 or 60 degrees in addition to 45 degrees, as illustrated in Figure 10.9, to avoid confusion. In large sections, only the borders need to be lined.

Figure 10.9 also illustrates that when a thin component such as a shim, gasket, or sheet-metal part is drawn in section, it is shown as a solid black area. Such components are drawn in this manner because they represent sections too small to be crosshatched.

10.4 TYPES OF SECTIONAL VIEWS

Several types of sectional views are utilized by the drafter to better clarify the interiors of objects and to assist in the ease of the production of mechanical parts, subassemblies, and assemblies. The most common sectional views include the full section; the half section; the broken-out section; the revolved section; the offset section; the removed section; sections through webs, spokes, and ribs; and breaks in elongated objects.

10.4.1 The Full Section

The **full section** is formed when the cutting plane line passes completely through an object, leaving one-

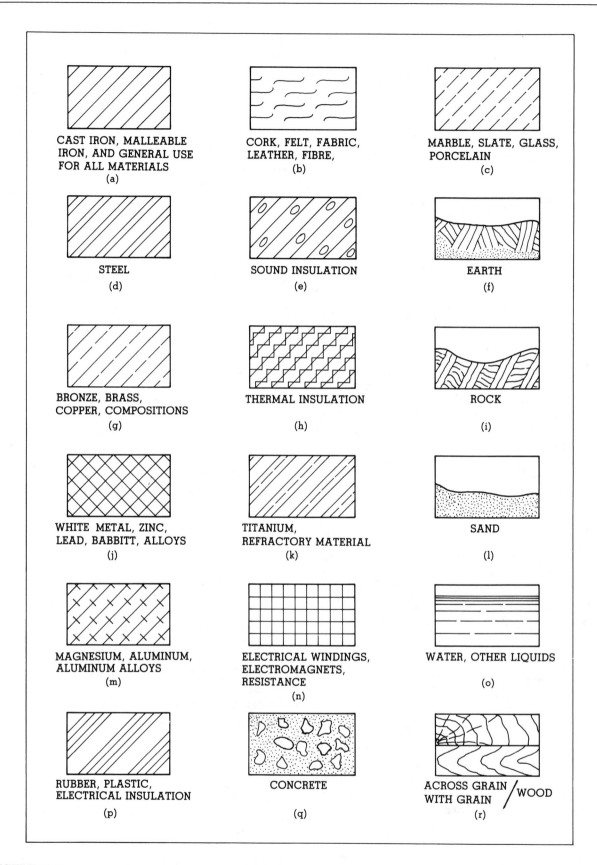

FIGURE 10.5
Standard code symbols for section lining (ANSI Y14.2).

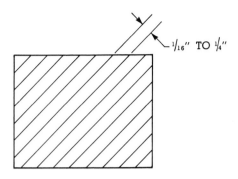

FIGURE 10.6
General-purpose section lining.

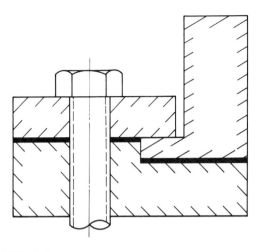

FIGURE 10.9
Section lining for three or more adjacent parts.

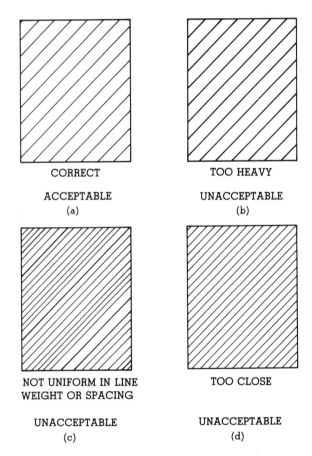

CORRECT
ACCEPTABLE
(a)

TOO HEAVY
UNACCEPTABLE
(b)

NOT UNIFORM IN LINE
WEIGHT OR SPACING
UNACCEPTABLE
(c)

TOO CLOSE
UNACCEPTABLE
(d)

FIGURE 10.7
Acceptable and unacceptable section lining practices.

FIGURE 10.8
Section lining or crosshatching of mating or adjacent parts.

half of the object exposed to the observer for viewing of its internal details. Figure 10.10(a) shows the top view of a support housing bracket. Figure 10.10(b) is the front view and identifies the internal detail of the object when the full sectional view A-A is taken. Note that the cutting plane line passes through the entire length of the object. The letters A-A represent the direction for viewing the section, and section lining is shown only where the cutting plane cuts through the object.

If sectional view B-B is desired, it would be represented correctly as in Figure 10.10(c), and could be shown as the right side view of the object.

The cutting plane line B-B cuts through the width of the object and indicates the viewing direction. Hidden lines are not usually shown in the sectional view.

Another example of a full section is illustrated in the hub in Figure 10.11. Note the detail shown in section A-A, which would not be clearly revealed if the right side were not shown in full section as in Figure 10.12.

10.4.2 The Half Section

The **half section** is usually a view of a symmetrical object which shows both internal and external fea-

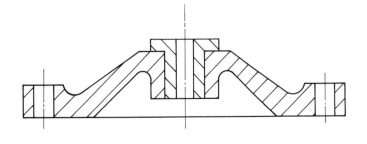

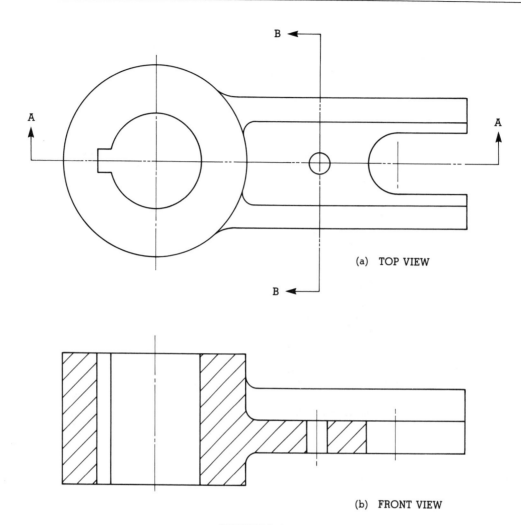

(a) TOP VIEW

(b) FRONT VIEW

SECTION A-A

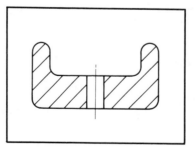

RIGHT SIDE VIEW

SECTION B - B

(c)

FIGURE 10.10

(a) and (b) Full sectional view of support housing bracket; (c) Right side view of support housing bracket in full section.

FIGURE 10.11
Right side view of hub in full
section.

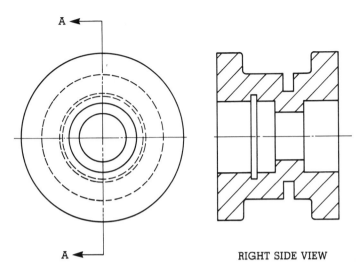

RIGHT SIDE VIEW

SECTION A - A

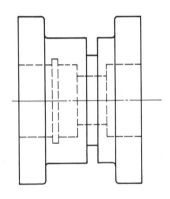

RIGHT SIDE VIEW

FIGURE 10.12
Right side view of hub not shown in section.

tures. This is done by passing two cutting planes at
right angles to each other along the center lines of
symmetrical axes so that one-quarter of the object is
removed and the interior detail is exposed to the
viewer. Examples of cutting planes at right angles to
each other are illustrated in Figure 10.13.

When a half section of a symmetrical object is
required, as shown in Figure 10.14, one-fourth of
the object is removed.

10.4.3 The Broken-out Section

When a sectional view of only a portion of an object
is required, the **broken-out section** may be utilized.
An irregular break line is normally used to determine
the extent of the section. Neither a cutting plane line
nor letters (A-A) to identify the section are required.
Examples of broken-out sections are illustrated in
Figure 10.15.

10.4.4 The Revolved Section

The **revolved section** provides a means of illustrat-
ing the shape of the cross section of an object and is
drawn directly on the exterior view as depicted in
Figure 10.16. This figure is a partial view of a wheel
showing three different revolved sections. Note that
the cutting plane is passed perpendicular to the center line or axis of the part to be sectioned and the
resulting section is rotated in place. When the object's visible lines interfere with the section, the view

FIGURE 10.13
Cutting planes at right angles
to each other.

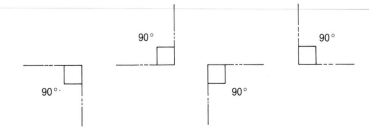

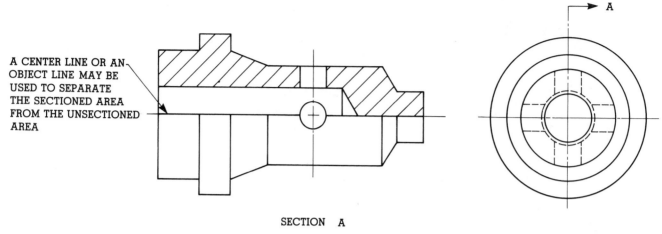

A CENTER LINE OR AN OBJECT LINE MAY BE USED TO SEPARATE THE SECTIONED AREA FROM THE UNSECTIONED AREA

SECTION A

FIGURE 10.14
Half section of a symmetrical object.

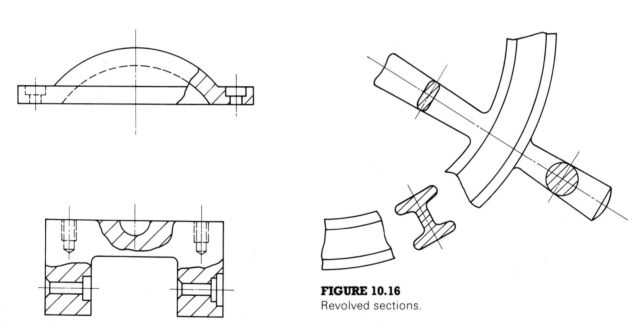

FIGURE 10.16
Revolved sections.

FIGURE 10.15
Broken-out sections.

is broken away by an irregular line to allow a clear space for the sectional views, and the cutting plane indications are omitted.

10.4.5 The Offset Section

At times it is necessary to draw the cutting plane line in a direction other than that of the main axis in order to show internal features which are not positioned in a straight line. When the cutting plane is not a single, continuous plane, the resulting section is called an

offset section. Figure 10.17 identifies the offset section A-A of a housing whose cutting plane is not a straight line but shows the interior of the object very clearly because of its offset nature. The section lining of the object is shown as if the offset were not present. Therefore, the offset section cutting plane must always be identified on the drawing, since the offset cannot be detected in the sectional view.

10.4.6 The Removed Section

The **removed section** may be used to illustrate specific areas or parts of an object. It is drawn like a revolved section, as demonstrated in Figure 10.18, except that it is placed to one side of the object. This

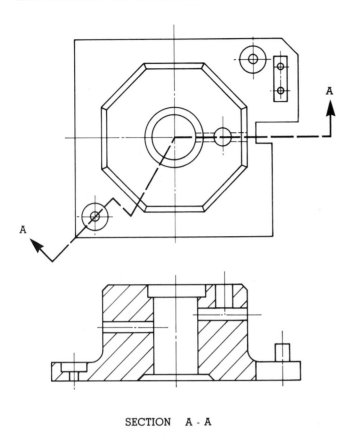

SECTION   A - A

FIGURE 10.17
Offset section.

figure shows how the removed section is drawn as well as the difference between the removed and revolved sections. Note, too, that cutting plane lines are not shown in the section. This type of section is often drawn to a larger scale than that of the view in which it is indicated, to show important data in greater detail. When this occurs, the scale of the section is noted directly below the section, as shown in Figure 10.18.

10.4.7 Sections Through Webs, Spokes, and Ribs

Sometimes conventional technical drawing practices may be violated for the sake of clarity because true projection of a sectional view often creates incorrect impressions of the actual shape of an object and may mislead the observer into thinking the object is shaped in a certain manner when, in fact, it is not. For this reason sectional views through webs, spokes, and ribs are treated in a special way.

10.4.7.1 Webs and Spokes

A **web** is an interior piece of metal plate or sheet which is used to connect heavier sections of an object. A **spoke** is a rod, finger, or brace that connects the hub with the rim of a wheel. The pictorial view in Figure 10.19(a) shows a pulley whose interior con-

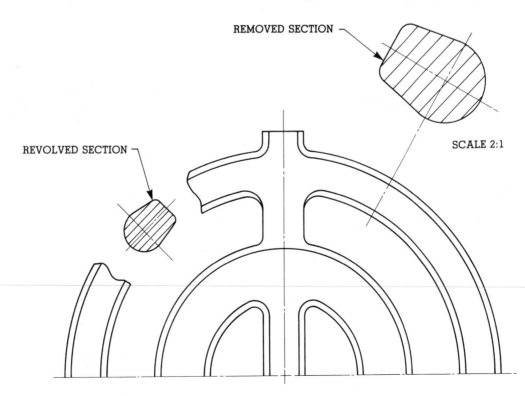

REMOVED SECTION

REVOLVED SECTION

SCALE 2:1

FIGURE 10.18
Removed and revolved sections.

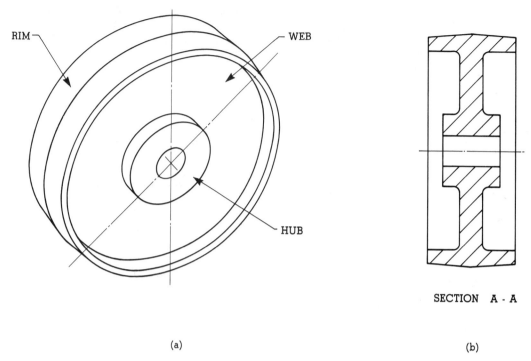

(a)   (b)

FIGURE 10.19
(a) Pulley; (b) Full section of web.

necting section (web) is attached to the hub (center) and the rim (outer section) of the pulley.

The solid web shown in Figure 10.19(a) is drawn in full section, in accordance with conventional drawing practices, in Figure 10.19(b). However, if the pulley had four spokes instead of a solid web, its full sectional view would be as shown in Figure 10.20. Even though the cutting plane passes directly through two of the spokes, the sectional view of the figure must be represented without section lining in order to avoid the appearance of a solid web.

An additional example of spokes in section is illustrated in Figure 10.21. Figure 10.21(a) is an unacceptable, confused sectional view that results when the pulley is drawn in true projection. Figure 10.21(b) is not acceptable because it portrays the pulley as having a solid web, which it does not. The only acceptable section of the pulley is Figure 10.21(c), which preserves the symmetry by showing the spokes as if they were aligned in a single plane.

10.4.7.2 Ribs

A **rib** is a part or a piece which serves to shape, support, or strengthen an object. A pictorial example of a bearing support with ribs is shown in Figure 10.22.

The true, but unacceptable, projection of section A-A of the bearing support is illustrated in Figure 10.23. This projection is unacceptable because it is

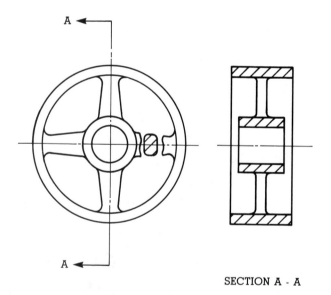

FIGURE 10.20
Correct method of showing spokes in section.

misleading to the observer. The ribs are not identified.

The conventional practice for sectioning of an object with ribs is to draw the section as represented in Figure 10.24, which is not a true projection but is the acceptable method. Note that there is no section lining for the ribs.

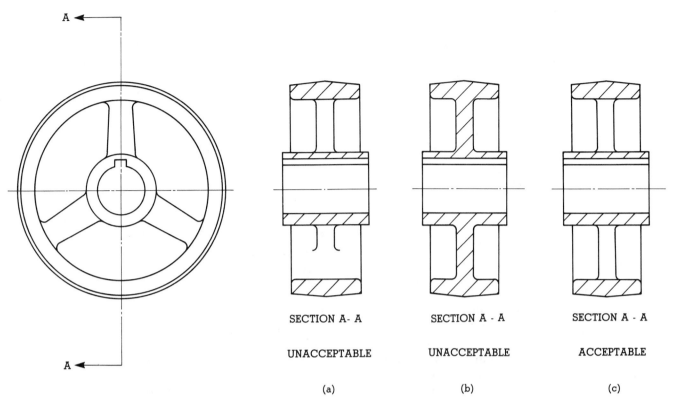

SECTION A- A   SECTION A - A   SECTION A - A

UNACCEPTABLE   UNACCEPTABLE   ACCEPTABLE

(a)   (b)   (c)

FIGURE 10.21
Conventional representation of spokes in section.

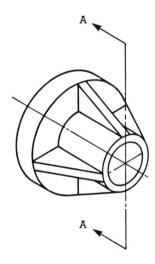

FIGURE 10.22
Pictorial view of a bearing support with ribs.

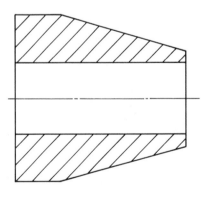

SECTION A - A

FIGURE 10.23
Unacceptable true projection of a bearing support with ribs.

An acceptable alternative method of presentation is to portray the line of intersection of the rib with a hidden line, as shown in Figure 10.25. In crosshatching the rib, every other section line is omitted to emphasize the division of the rib and the body of the object.

10.4.8 The Unlined Section

The **unlined section** is used primarily for items which are standard, "off-the-shelf," purchased parts or items of hardware. Shafts, nuts, bolts, rods, rivets, keys, pins, ball bearings, set screws, and other types

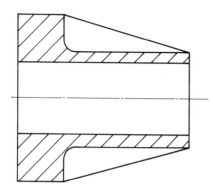

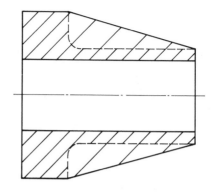

SECTION A - A

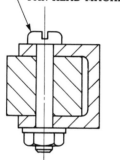

SECTION A - A

FIGURE 10.24
Acceptable sectioning practice for objects with ribs.

FIGURE 10.25
Alternative acceptable sectioning method for objects with ribs.

of fasteners whose axes lie in the cutting plane are not sectioned. Application of section lining to such components, which have no internal detail, would serve no useful purpose and in fact would tend to confuse the observer. Figure 10.26 provides several examples of items which do not require cross-hatching.

10.5 BREAKS IN ELONGATED OBJECTS

Normally, elongated objects such as solid shafts, bars, tubing, and wood sections cannot be drawn to their entire lengths on a drawing medium. Therefore, in drawing an object of this type a **break** located at a

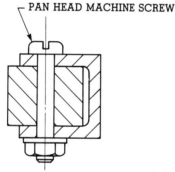

PAN HEAD MACHINE SCREW

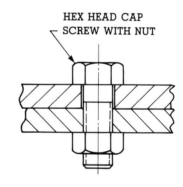

HEX HEAD CAP
SCREW WITH NUT

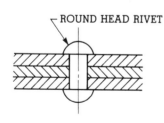

ROUND HEAD RIVET

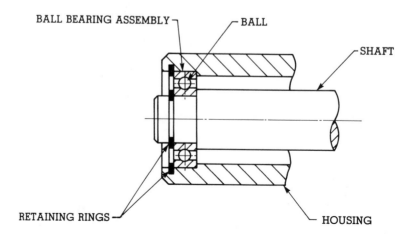

BALL BEARING ASSEMBLY   BALL

SHAFT

RETAINING RINGS

HOUSING

FIGURE 10.26
Items not sectioned.

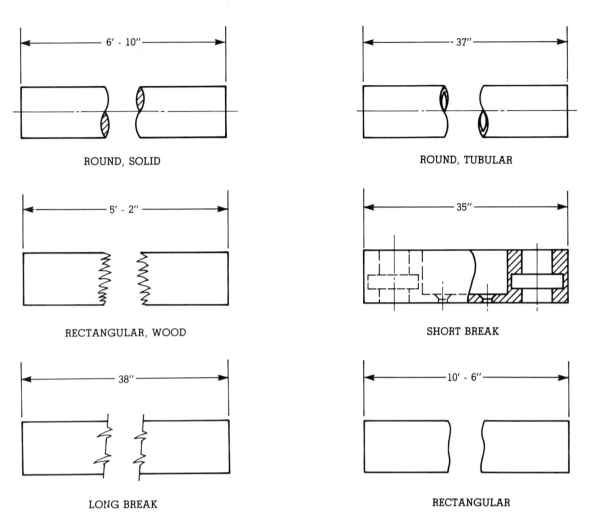

FIGURE 10.27
Conventional breaks used in drawing elongated parts.

convenient position is used and the true length is indicated by a dimension. Very often an object can be drawn to a larger scale to improve legibility if a break is used. Examples of conventional breaks are demonstrated in Figure 10.27. Breaks can be produced freehand or with a template, compass, or irregular curve.

10.6 SUGGESTIONS FOR PRODUCING THE SECTIONAL VIEW

There are several points to keep in mind when producing sectional views. The following basic guidelines and suggestions should be practiced if quality sectional views are to be produced.

1. When practical, a sectional view should be shown in the correct orthographical projection with other views.

2. The cutting plane line should be drawn so that it stands out clearly and cannot be confused with other lines.
3. The cutting plane line may be omitted when it corresponds with the center line of an object and if it is obvious where the cutting plane lies.
4. The general-purpose section lining symbol should be used on all drawings unless a specific, required application is needed.
5. Only the borders of large areas need to have section lining.
6. In sections through webs, spokes, and ribs, conventional technical drawing practices may be violated for the sake of clarity, if true projection creates an incorrect impression of the actual shape of the object.
7. Items of hardware or parts whose axes lie in the cutting plane should not be sectioned.
8. When three or more adjacent parts are shown in section, or when an assembly drawing is shown

in section, section lining should be drawn at 30 or 60 degrees in addition to 45 degrees to avoid confusion.

9. Use consecutively lettered sections A-A, B-B, C-C, etc., to identify the sectional view. Arrange them in alphabetical order from left to right.

10. Section lining should not cross dimension, extension, or leader lines or obscure other important details on the sectional view.

11. Invisible lines need not be shown in the sectional view.

12. If the scale of a sectional view is different from that of the basic drawing, it should be identified as such.

10.7 SUMMARY

The purpose of the sectional view is to clearly show the internal form and detail of an object whose interior construction cannot be easily identified by an exterior view. The sectional view is obtained by passing an imaginary line, called the cutting plane line, through some specific part of the object and presenting the object as though the part cut off by this line had been removed. Three different acceptable cutting plane lines were introduced. Conventions for sectioning were outlined and several approved symbols were identified. Acceptable and unacceptable section lining practices were presented.

The most commonly utilized sectional views were described and covered in detail, including the full section; the half section; the broken-out section; the revolved section; the offset section; the removed section; sections through webs, spokes, and ribs; and breaks in elongated parts. The various sectional views and their associated technical terms were defined, and examples of each were shown in detail. Acceptable and unacceptable sectioning practices were provided to help the learner internalize correct procedures. The uses of unlined sections and breaks in elongated objects were discussed, and examples of each were presented. Several basic but important guidelines were presented, and suggestions were offered for producing the sectional view. Practice exercises for this section are presented on pages 10A through 10I in the workbook.

10.8 **REVIEW EXERCISES**

1. What is the purpose of the sectional view?

2. Of what use is the cutting plane line?

3. Sketch three examples of cutting plane lines.
  a.

  b.

  c.

4. In the view below, sketch the general-purpose section lining symbol.

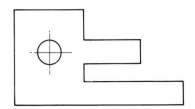

5. Apply section lining to the drawing of assembled parts below.

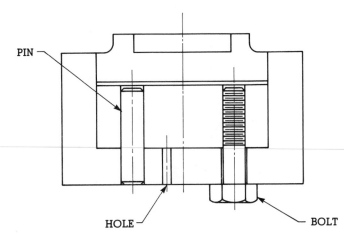

6. List six kinds of sectional views.

   a.

   b.

   c.

   d.

   e.

   f.

7. In a full section, how much of the object is removed?

8. In a half section, how much of the object is removed?

9. What kind of line is used to identify the broken-out section?

10. What kind of sectional view is recommended when the cutting plane line lies in a direction other than that of the main axis?

11. Explain why webs, spokes, and ribs are not crosshatched.

12. What kind of sectional view identifies both the internal details and external features of an object at the same time?

13. List eight items whose axes lie in the cutting plane which are not section lined.

    a.

    b.

    c.

    d.

    e.

    f.

    g.

    h.

14. What are two advantages of producing breaks in elongated objects?

15. What do the arrows on a cutting plane line indicate?

16. List eight basic guidelines and suggestions for producing sectional views.

    a.

    b.

    c.

    d.

    e.

    f.

    g.

    h.

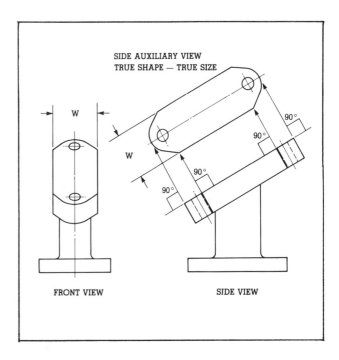

SIDE AUXILIARY VIEW
TRUE SHAPE — TRUE SIZE

W

W

90°

90°

90°

90°

FRONT VIEW

SIDE VIEW

Section 11

The Auxiliary View

LEARNER OUTCOMES

The student will be able to:

- Identify the need for producing an auxiliary view

- List four basic types of auxiliary views

- Name three kinds of primary auxiliary views

- Explain the steps required to construct a secondary auxiliary view

- Describe the procedure for producing a curved surface auxiliary view

- Identify several important suggestions for producing an auxiliary view

- Demonstrate learning through successful completion of review and practice exercises

11.1 PURPOSE

Sometimes objects have inclined, sloped, or slanted surfaces that are not parallel with any of the three principal planes of projection. Figure 11.1 illustrates such an object. When this object is drawn in orthographic projection, as in Figure 11.2, in the top and side views the inclined surface "A" appears to be **foreshortened**, and the true shape of this surface is not apparent.

To truly represent the inclined surface and correctly illustrate its true shape, an **auxiliary view** should be produced. Figure 11.3 not only shows the three orthographic views but also identifies the true shape of surface "A" through the addition of an auxiliary view. An auxiliary view is developed as though the **auxiliary plane** were hinged to the plane of the object to which it is perpendicular and then revolved into the plane of the paper.

The purpose of the auxiliary view is threefold: (1) to present to the observer a line of sight of an object that is perpendicular to the inclined surface, as viewed by looking directly at the inclined surface; (2) to illustrate the true shape and form of a surface; and (3) to project and complete other views of an object. Only the true shape of the inclined surface needs to be drawn. No additional features of the orthographic views are required. Hidden lines are normally omitted in the auxiliary view unless they are required for clarity.

11.2 TYPES OF AUXILIARY VIEWS

Several different kinds of auxiliary views are produced by the drafter depending on the complexity of the object and the type of drawing required. Four basic types—the primary, secondary, partial, and curved surface auxiliary views—will be covered in this section.

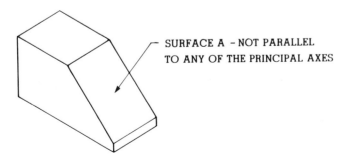

FIGURE 11.1
Object with inclined surface.

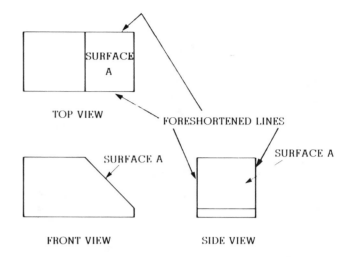

FIGURE 11.2
Orthographic projection of object in Figure 11.1.

11.2.1 The Primary Auxiliary View

Auxiliary views may assume many different positions in relation to the three principal planes of projection. The **primary auxiliary view** may be categorized into three general types, based on their positions relative to the principal (front, top, and side) views. Figure 11.4 illustrates the **front auxiliary view**, where the auxiliary plane is perpendicular to the front view, called the **frontal plane**, and inclined to the horizontal plane of the projection. Note that the auxiliary view and the top view have one dimension in common—the width (W) of the object. The front auxiliary view is drawn as if it were hinged to the frontal plane which is projected from the front view.

The **top auxiliary view** is one in which the inclined surface is perpendicular to and projected from the top view, whose height (H) is the same as the height in the front view. This type of view is presented in Figure 11.5.

The **side auxiliary view** is portrayed in Figure 11.6. This type of auxiliary view is projected from the side view and has a common width (W) dimension with the front view. To construct it, measurements may be taken from the front view.

All three types of primary auxiliary views are constructed in similar fashion. Each is projected from the view in which the inclined surface appears as a line, and each has its measurements taken from the principal view with which it has a common dimension. Each of the three, when drawn, is considered to be hinged to the principal plane to which it is perpendicular.

The general procedure for constructing the primary auxiliary view is outlined below and illustrated in Figure 11.7.

FIGURE 11.3
Orthographic projection with
auxiliary view.

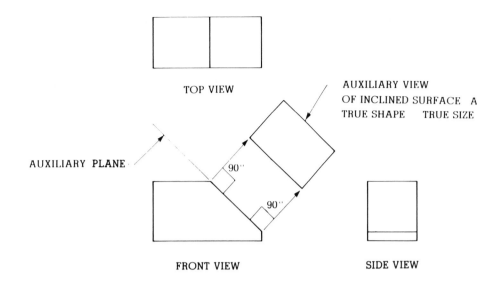

FIGURE 11.4
The front auxiliary view.

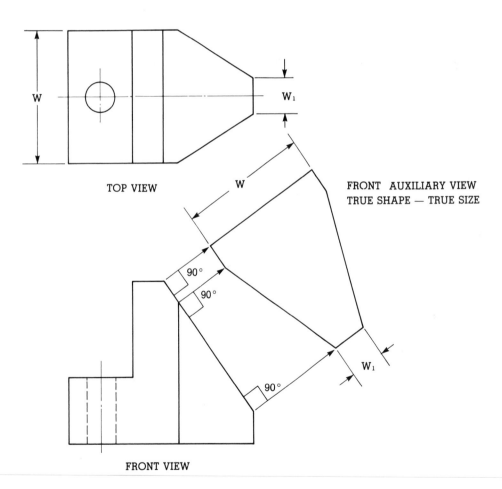

Given: A pictorial view of a channel clip (see Figure
11.7). Because the inclined surface is part of the
front view, a front auxiliary view will be constructed.

Step 1: Draw two related principal views of the
channel clip, in this case the front and right side
views. Label all points which will affect the construction of the auxiliary view. In the right side
view, draw a light **reference line** vertically as an
extension of AC.

Step 2: Project light lines at right angles (90 degrees) from the ends of the inclined surface. Place
reference line (AC) parallel with and at any convenient distance from the inclined surface.

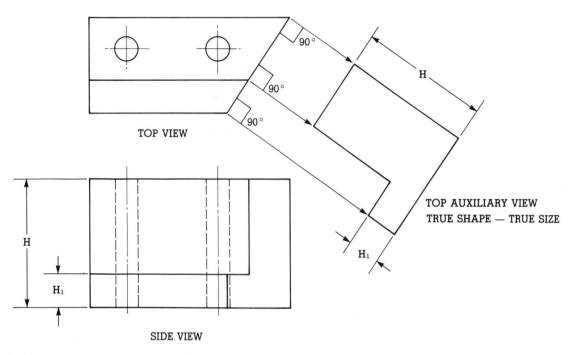

FIGURE 11.5
The top auxiliary view.

FIGURE 11.6
The side auxiliary view.

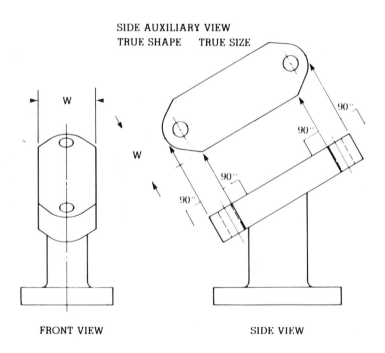

Step 3: From the right side view, take measurements to each lettered point directly from the reference line AC. Using dividers, as shown in the illustration, is the preferred method for transferring measurements to the reference line of the inclined surface. Label each point with the appropriate letter (A through H).

Step 4: When satisfied that all measurements have been transferred accurately and the construction is complete, darken all lines in all views. Erase all letters (A through H) and projection and construction lines. The construction of the front auxiliary view, which is a primary auxiliary view, is now complete. Note that all of the inclined surface has

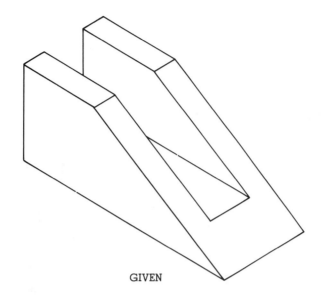

GIVEN

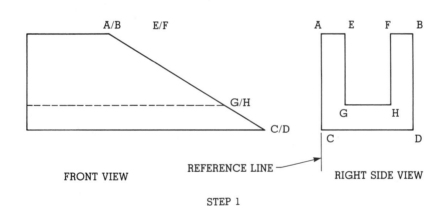

A/B   E/F

FRONT VIEW

G/H

C/D

A   E   F   B

G   H

C   D

REFERENCE LINE

RIGHT SIDE VIEW

STEP 1

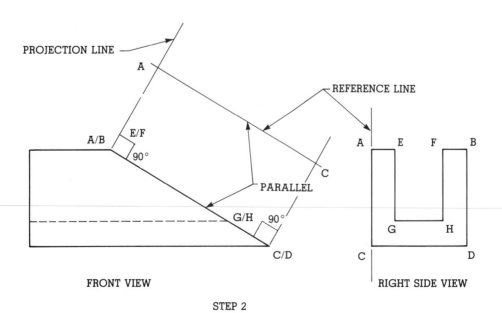

PROJECTION LINE

A

A/B   E/F

90°

G/H   90°

C/D

PARALLEL

C

REFERENCE LINE

A   E   F   B

G   H

C   D

FRONT VIEW

RIGHT SIDE VIEW

STEP 2

FIGURE 11.7
Procedure for constructing a primary auxiliary view of a channel clip. (Continued on page 150.)

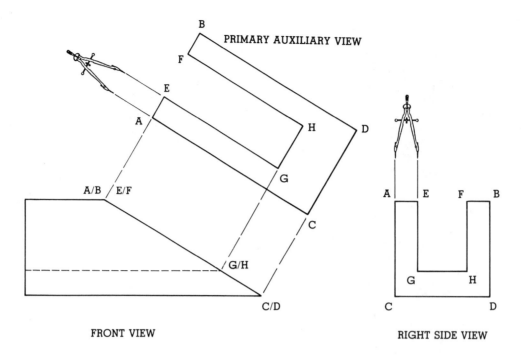

FRONT VIEW   RIGHT SIDE VIEW

STEP 3

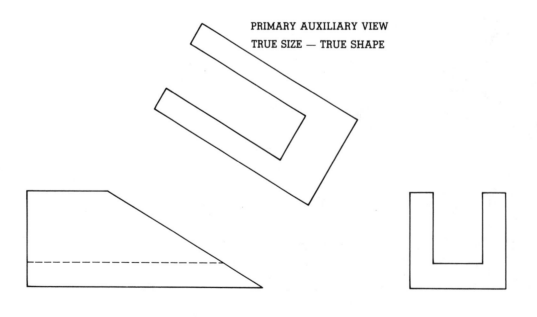

FRONT VIEW   RIGHT SIDE VIEW

STEP 4

FIGURE 11.7 (continued)

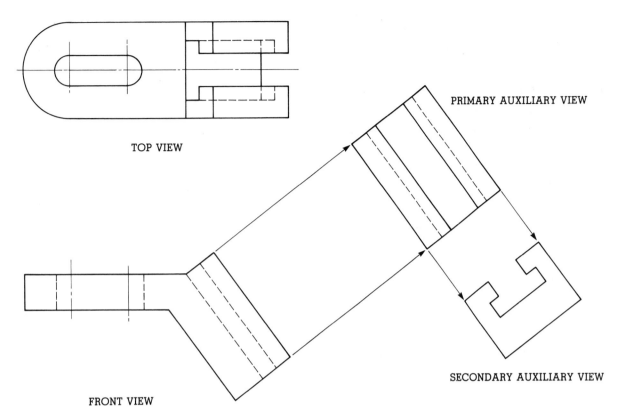

TOP VIEW

PRIMARY AUXILIARY VIEW

FRONT VIEW

SECONDARY AUXILIARY VIEW

FIGURE 11.8
Secondary auxiliary view of a support bracket.

been projected into the auxiliary view, which shows its true size and true shape.

11.2.2 The Secondary Auxiliary View

At times, the primary auxiliary view will not sufficiently illustrate all the detail of an object and, therefore, a **secondary auxiliary view** may be required. The secondary auxiliary view may be defined as a supplementary view that has been projected directly from a primary auxiliary view.

In order to construct a secondary auxiliary view, the primary auxiliary view is constructed first so that it is perpendicular to the inclined surface and to one of the principal planes, as shown in Figure 11.8. In this example, the primary auxiliary view is constructed from the front view of the support bracket and the secondary auxiliary view is produced from the primary auxiliary view. The same general procedure used to construct the primary auxiliary view is used for producing the secondary auxiliary view.

11.2.3 The Partial Auxiliary View

Another frequently used supplementary view is called the **partial auxiliary view**. This type of view tends to simplify the drawing of an object, shortens the draft-

ing time, and makes a drawing easier to read. It allows the drawing to be more functional without sacrificing clarity or quality, and its omitted features are shown in another view. The mounting bracket in Figure 11.9 is shown with a complete side view, a partial front view, and a partial auxiliary view. Break lines are used at convenient locations in the partial views.

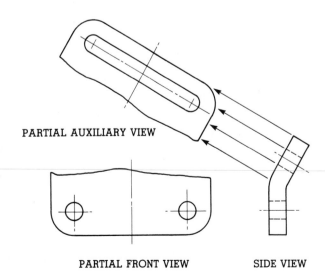

PARTIAL AUXILIARY VIEW

PARTIAL FRONT VIEW

SIDE VIEW

FIGURE 11.9
Partial auxiliary and front views of a mounting bracket.

The construction process for producing partial views follows the same general guidelines as those for the primary and secondary auxiliary views.

11.2.4 The Curved Surface Auxiliary View

If an object has circular features such as the **truncated** or **beveled** end of the tube section shown in the pictorial in Figure 11.10, then a **curved surface auxiliary view** should be drawn to show its true shape and size. The method most commonly used to construct the auxiliary view of a curved surface is to plot a series of points and then connect them with

a line. The number of points will determine the smoothness and accuracy of the desired curve. When the true size is obtained it will be in the form of an ellipse. The process for producing a curved surface auxiliary view is presented in the sequential steps that follow and is illustrated in Figure 11.10.

Given: A pictorial view of a tube section (see Figure 11.10).

Step 1: Draw two related principal views of the tube section—the front and right side views. Also include light construction lines for the top view. Divide the right side view into 12 equal parts. Starting at the top and moving clockwise, label the

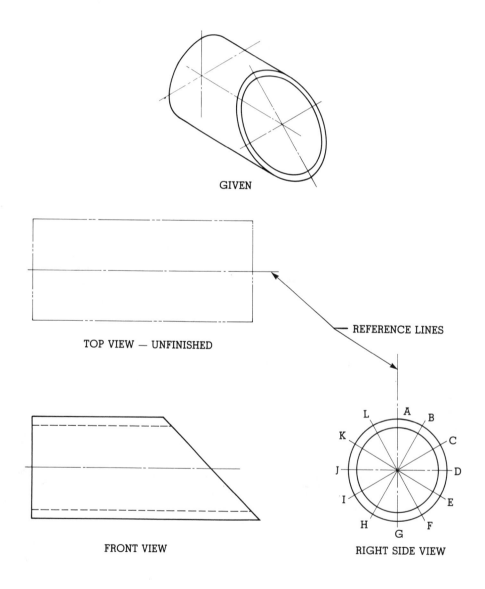

GIVEN

TOP VIEW — UNFINISHED

— REFERENCE LINES

FRONT VIEW

RIGHT SIDE VIEW

STEP 1

FIGURE 11.10
Procedure for constructing a curved surface auxiliary view of a tube section. (Continued on pages 153 and 154.)

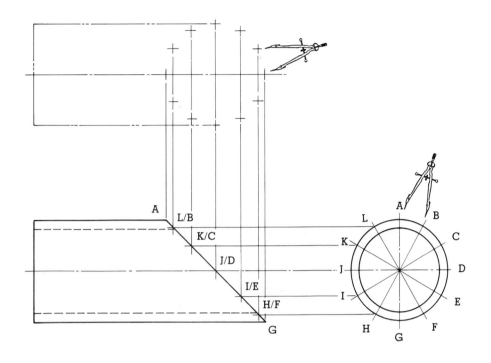

STEP 2

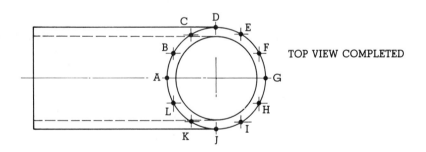

TOP VIEW COMPLETED

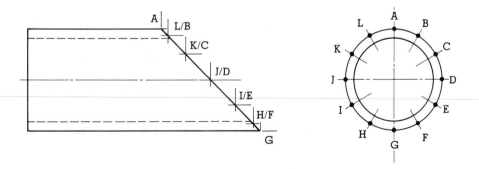

STEP 3

FIGURE 11.10 (continued)

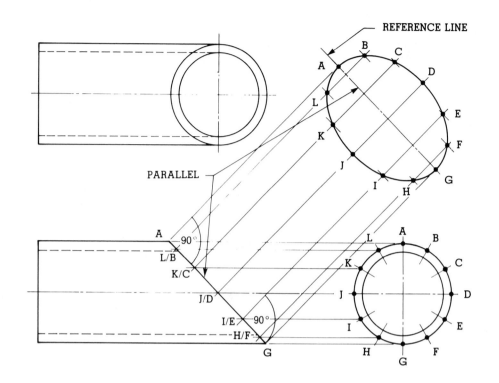

STEP 4

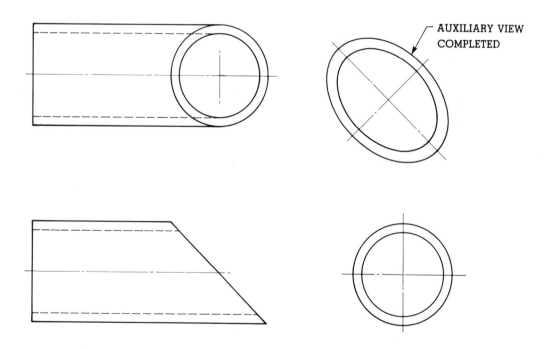

STEP 5

FIGURE 11.10 (continued)

points of intersection with the letters A through L. Vertical center line AG will be used as a reference line in the right side view, and the horizontal center line is also a reference line in the unfinished top view.

Step 2: Project light horizontal lines from the 12 established points (A through L) in the right side view to the inclined edge in the front view. Draw light projection lines vertically from the intersected inclined plane of the front view to the top view. Using dividers, transfer measurements from the right side view to the top view and label each. Take each measurement from the vertical reference line AG.

Step 3: Connect all points using light-weight lines and an irregular curve. When satisfied that the top view curved surface is as smooth as possible, "heavy up" all lines. The top view is now complete.

Step 4: Draw light projection lines at right angles (90 degrees) to the inclined surface in the front view. At any convenient distance, place the reference line AG parallel to the incline. Project lines from points A, L/B, K/C, J/D, I/E, H/F, and G in the front view to the auxiliary view location. As previously performed in step 2, transfer measurements from the right side view to the auxiliary view reference line using dividers. Label each point of intersection.

Step 5: Connect all points on the auxiliary view using light-weight lines and an irregular curve. Erase all letters and construction lines. When satisfied that the ellipse-shape auxiliary view is as smooth as possible, "heavy up" the lines. The curved surface auxiliary view is now complete.

11.3 SUGGESTIONS FOR PRODUCING THE AUXILIARY VIEW

The following are important points to consider when constructing an auxiliary view.

1. Allow sufficient space for all views.

2. An auxiliary view is developed as though the auxiliary plane were hinged to the plane to which it is perpendicular.

3. Only the true shape features of the inclined surface need to be drawn unless otherwise directed.

4. Hidden lines are normally omitted in the auxiliary view unless required for clarity.

5. At least two principal views are required to produce an auxiliary view.

6. Use dividers to transfer measurements from view to view.

7. Label all points which will affect the construction of the drawing.

8. A secondary auxiliary view is produced from a primary auxiliary view.

9. When constructing a curved surface auxiliary view, draw the ellipse-shape view as smooth as possible using an irregular curve.

11.4 SUMMARY

In order to show the true size and shape of an object which has an inclined surface, an auxiliary view of the inclined surface should be produced. The auxiliary view is developed as though it were hinged to the surface to which it is perpendicular. The purpose of the auxiliary view was explained as being threefold: (1) to present to the observer a line of sight of an object that is perpendicular to the inclined surface, as viewed by looking directly at the inclined surface; (2) to illustrate the true shape and form of a surface; and (3) to project and complete other views of an object.

Four basic types of auxiliary views were covered—the primary, secondary, partial, and curved surface auxiliary views. As subsets of the primary auxiliary view, the front, top, and side auxiliary views were detailed. A step-by-step procedure for producing each of the basic types of views was presented, and suggestions for producing an auxiliary view were outlined. Practice exercises for this section are presented on pages 11A through 11H in the workbook.

11.5 REVIEW EXERCISES

1. What is the purpose of the auxiliary view?

2. List the four basic types of auxiliary views.

3. Name the three kinds of primary auxiliary views.

4. Complete the following:
   a. The top auxiliary view is one in which the inclined surface is perpendicular to and projected from the _____ view.
   b. The _____ auxiliary view is one in which the inclined surface is perpendicular to and projected from the front view.

5. What must be done before a secondary auxiliary view can be constructed?

6. What type of auxiliary view simplifies the drawing of an object, shortens the drafting time, and makes a drawing easier to read?

7. At what angle should projection lines be drawn from the inclined surface of an object?

8. What is the purpose of a reference line in the construction of an auxiliary view? Where should it be placed? At what angle or position?

9. When are hidden lines used in an auxiliary view?

10. When would a curved surface auxiliary view be required?

11. List six important points to consider when producing an auxiliary view.

    a.

    b.

    c.

    d.

    e.

    f.

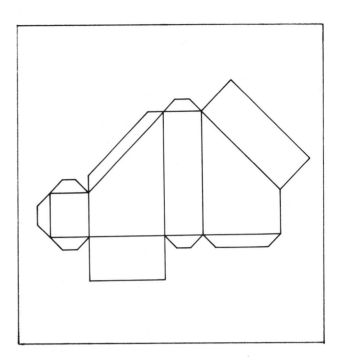

Section **12**

The Development Drawing

LEARNER OUTCOMES

The student will be able to:

- Identify the major types of development drawings

- Describe the characteristics of the parallel line development, the radial line development, and the triangulation development

- Define the purpose of the fold line, the base line, the seam line, and tabs in the construction of development drawings

- State the purpose of the development drawing

- Define the function of the true-length diagram

- Describe the intent of the transition piece

- Produce several examples of development drawings

- Demonstrate learning through successful completion of review and practice exercises

12.1 PURPOSE

The **development drawing** is usually a design layout of a **pattern** or a **template**. It is drawn and developed in a single plane by the drafter in preparation for folding, rolling, or bending to form some predetermined shape. These flat patterns or shapes facilitate the cutting of a desired **blank** from sheet metal or paper products. The purpose of the development drawing is to produce a model of an item with a particular shape or form, regardless of whether one or millions of the same item are to be produced. Examples of items that require development drawings include heating and air conditioning ducts, furnace parts, mailboxes, and disposable items such as cartons and liquid and dry food containers. A one-of-a-kind item could be a special plenum or transition piece that attaches to the top of a furnace. Such a piece is illustrated in Figure 12.1. The bottom is square and the top is round, hence the name "transition piece." It is made of thin-gage sheet metal.

An item which is disposable and of which millions are used daily is the paper cup. Paper containers that are round, conical, square, rectangular, and oval are produced in hundreds of sizes and shapes, as depicted in Figure 12.2.

Before a paper container assumes its final shape, there needs to be a design layout of the container, called a development drawing. The development of the sidewall blank of a typical paper cup, when laid out flat in a single plane, has the shape demonstrated in Figure 12.3(a). Its bottom blank is shown in Figure 12.3(b).

Figure 12.4 demonstrates how valuable the development drawing can be as it sequences the assembly of a paper cup. This cup is produced by automatic machinery at a rate of up to 200 units per minute.

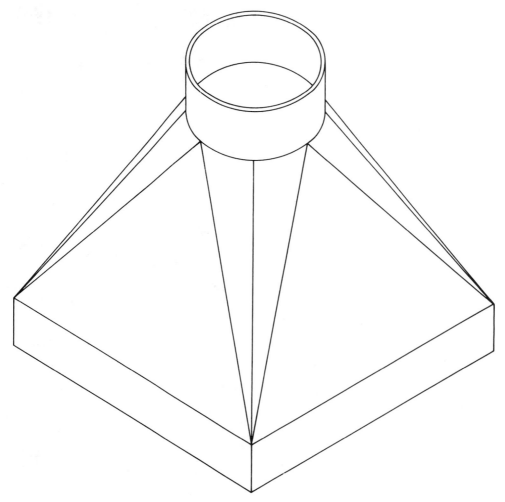

FIGURE 12.1
Furnace plenum or transition piece.

FIGURE 12.2
Paper cups and containers. (Courtesy of Paper Machinery Corporation).

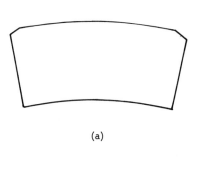

(a)

(b)

FIGURE 12.3
Sidewall and bottom blanks of a paper cup.

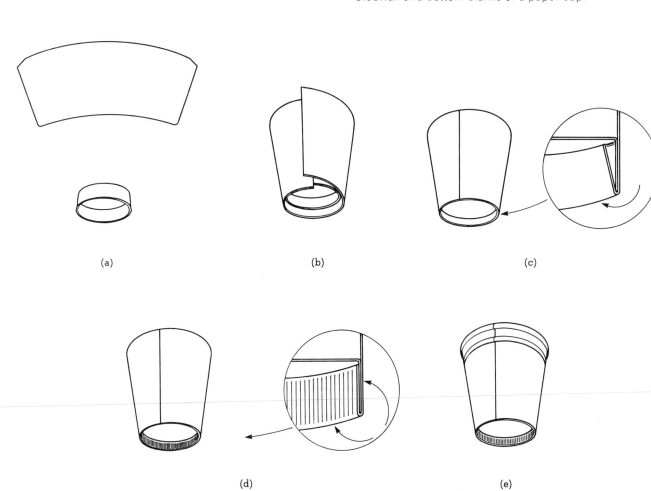

(a)   (b)   (c)

(d)   (e)

FIGURE 12.4
Sequence of operations for the assembly of a paper cup.

12.2 TYPES OF DEVELOPMENT DRAWINGS

A **surface** which can be unfolded or rolled out without distortion is **"developable,"** which means that it can be laid out flat in a single plane. Objects that consist of **single-curved surfaces** are considered to be developable. Examples of single-curved surfaces are presented in Figure 12.5.

Double-curved, distorted, or warped surfaces cannot be developed and, therefore, are termed **"nondevelopable."** These surfaces can be developed only by approximation and will appear in a development drawing as though the material has been stretched. For example, a sphere such as a ball cannot be wrapped smoothly because of its curvature, but if it is to be covered with a flexible, pliable, somewhat elastic material, the covering can be stretched to fit. Figure 12.6 provides examples of distorted or warped surfaces.

There are three major types of single-curved surface developments utilized in industry. They include the **parallel line development**, the **radial line development**, and the **triangulation development**. Double-curved surface developments are not covered in this textbook.

FIGURE 12.5
Developable, single-curved surfaces.

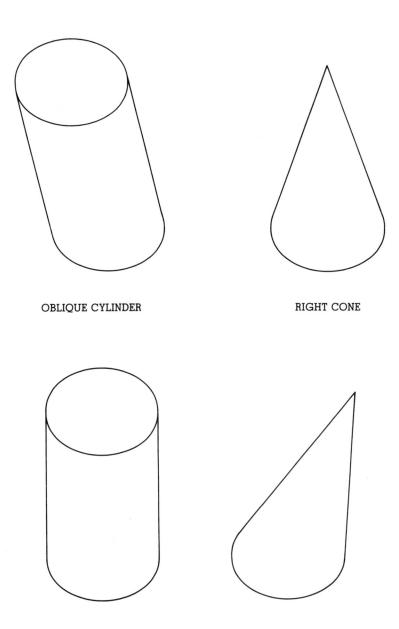

OBLIQUE CYLINDER

RIGHT CONE

RIGHT CYLINDER

OBLIQUE CONE

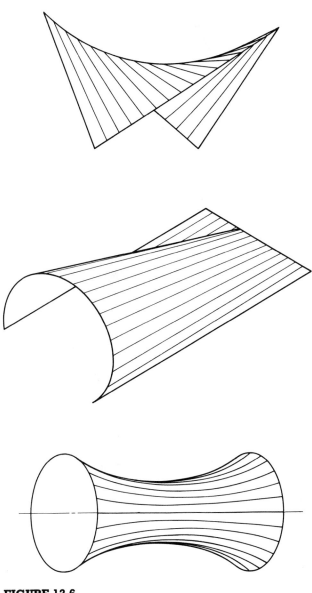

FIGURE 12.6
Examples of distorted or double-curved surfaces.

12.2.1 The Parallel Line Development

Prisms and rectangular, cubical, and cylindrical shapes are ones for which the **parallel line development** is utilized. Although a pictorial view is important for the visualization of an object, the development drawing requires orthographic projection of the views and surfaces involved.

Figure 12.7(a) is a pictorial view of a simple rectangular object whose material is thin-gage sheet metal. Note that all **lateral** surfaces are parallel. From the pictorial in Figure 12.7(a) an orthographic projection of three views is produced, as depicted in Figure 12.7(b). Two important technical terms need to be noted and defined at this point: the **base line**

and the **fold line**. The base line, also referred to as the **stretchout line**, is the point where the layout is developed or stretched out. Fold lines are lines which are parallel to each other and at 90 degrees to the base line.

If the object were unfolded it would appear as shown in Figure 12.7(c), and if it were stretched out or laid flat it would be as illustrated in Figure 12.7(d).

Because the object is made of thin-gage sheet metal, **tabs** are required to fasten the object together to complete its "box" shape, as represented in Figure 12.8. Tabs must be located so that each one attaches to an untabbed edge. Note that there are seven tabbed edges and seven untabbed edges. Each tab folds to attach to an untabbed edge. Width (W) of tabs will depend on the size of the object. Tabs are normally tapered at 45 degrees at the ends.

The step-by-step procedure for producing a development drawing of a **truncated prism** using parallel line development is outlined below and illustrated in Figure 12.9.

Given: A pictorial view of a truncated prism (see Figure 12.9).

Step 1: Scale the pictorial and draw front, top, right side, and front auxiliary views. Identify each view and the base line. Label each fold line.

Step 2: On a drawing medium such as vellum paper, draw a horizontal base line approximately 3 inches from the bottom of the paper to allow sufficient space for the complete development of the prism, including top and bottom surfaces and tabs. At a distance of 3½ inches from the left side, draw a vertical line perpendicular to the base line. This is the seam line where the development layout will start. Transfer vertical distance AB from the four-view drawing of step 1. Along the base line, transfer horizontal distances AC and BD. This locates the first fold line, which is parallel to line AB.

Step 3: From the base line, transfer horizontal distances DF, FH, and HB and vertical distances EF and GH. Return to line AB. Transfer distances for the top and bottom of the prism—namely, DB and FH for the bottom and AC and GE for the top.

Step 4: Draw fold lines CD, EF, GH, DF, and AG, as well as seam line AB. Connect all exterior lines until the outline is complete.

Step 5: Add a sufficient number of 3/16-inch-wide tabs to fasten the developed prism together. Make certain that the number of tabbed edges equals the number of untabbed edges. When satisfied that the layout is finished, erase all lettered points and "heavy up" all lines. This completes the parallel line development of the truncated prism.

On the print machine, make a copy of the truncated prism development drawing. Cut out the pat-

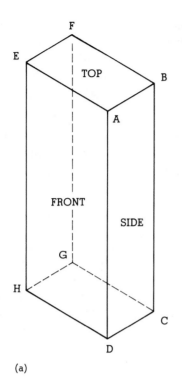

(a)

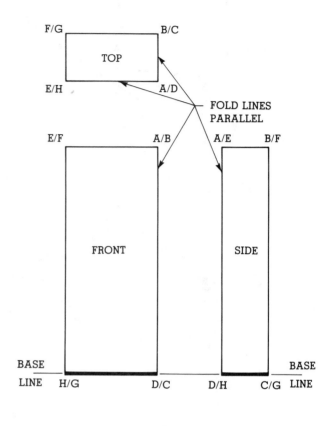

(b)

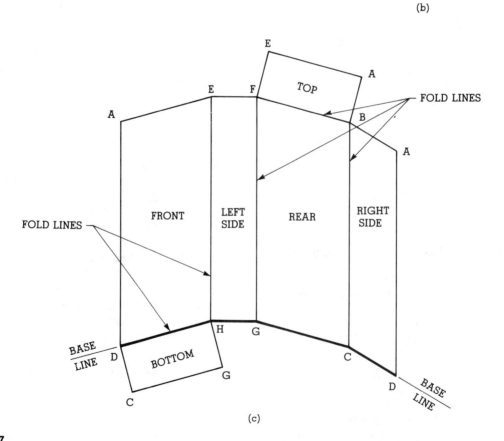

(c)

FIGURE 12.7

(a) Pictorial view of a rectangular object; (b) Orthographic views of object; (c) Unfolding object; (d) Stretched out or fully developed object. (Continued on page 165.)

FIGURE 12.7 (continued)

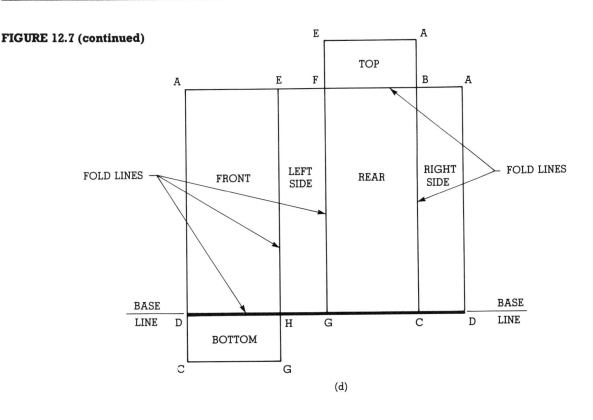

(d)

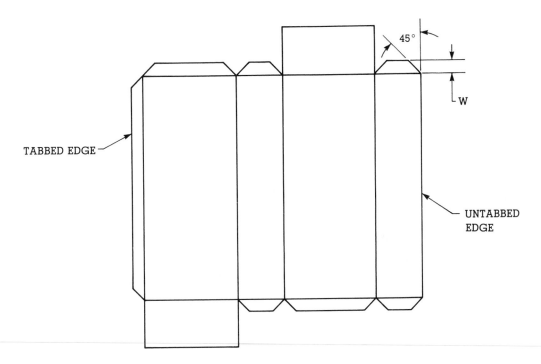

FIGURE 12.8
Tabbed edges.

tern. Fold at the fold lines as well as at tabbed lines. Fasten all tabbed areas with glue or tape until the prism is complete. Check to see if the completed truncated prism looks like the one in Figure 12.9.

The parallel line development of a **truncated cylinder** differs from that of a truncated prism. In the development of a round or partially round object, the fold line locations are selected from the **radial top**

FIGURE 12.9
Procedure for constructing a
development drawing of a
truncated prism.

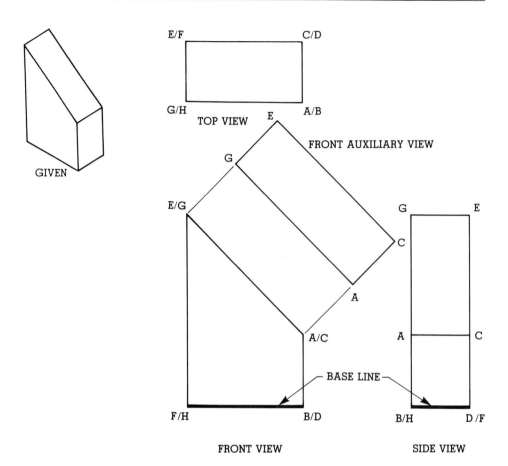

STEP 1

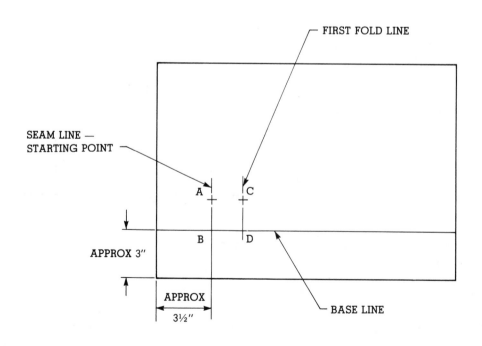

STEP 2

FIGURE 12.9 (continued)

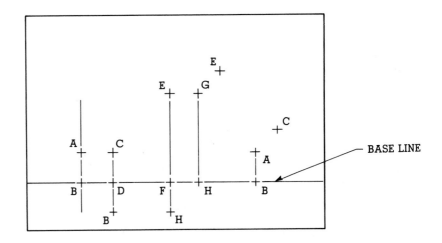

STEP 3

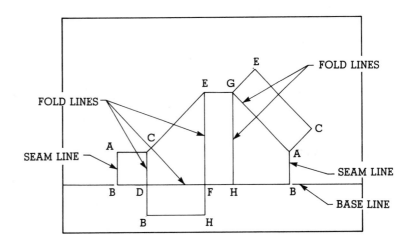

STEP 4

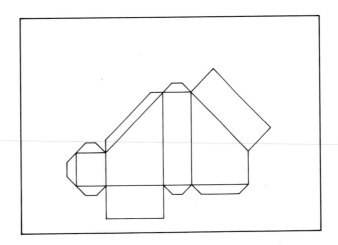

STEP 5

view of the object, and height locations are taken from the front view, as shown in Figure 12.10.

The procedure for producing a development drawing of a truncated cylinder using parallel line development is outlined as follows and is illustrated in Figure 12.11.

Given: A pictorial view of a truncated cylinder (see Figure 12.11).

Step 1: Draw front, top (radial), and front auxiliary views. Identify each view as well as the base line.

Step 2: So that the curved surface of the development will be as smooth as possible, divide the diameter (top view) of the cylinder into 12 equal parts. These divisions represent and locate the fold lines for the object and are needed for layout of the development. Identify the divisions by labeling them with letters A through L, beginning at the shortest height of the cylinder and proceeding clockwise. Project each of the lettered intersections downward to the front view to determine the **true lengths** of the fold lines. Fold lines are located 90 degrees from the base line.

Step 3: On a separate drawing medium, draw a horizontal base line 2½ inches from the bottom of the medium, allowing enough space for the complete development stretchout including the top and bottom of the cylinder. Three inches from the left side, draw a vertical seam line perpendicular to the base line. This will be the starting point for the development. Transfer height AA from the front view in step 2. There are two methods for determining the development stretchout: (1) by transferring **chordal** distances AB, BC, CD, DE, EF, FG, GH, HI,

IJ, JK, and KL from the top view and laying them out on the horizontal base line, using dividers; and (2) by the calculation method using the formula for the circumference of a circle, $C = 3.14 \times$ diameter. Either method is acceptable depending on the desired accuracy. Continue using the dividers for plotting heights. Transfer fold lines AB, CD, etc. until the rough outline of the development is visible. Note that the first vertical line and the final vertical line AA are of the same height.

Step 4: Add the bottom and the truncated top to the cylinder development, which will now take the form of the previously drawn front auxiliary view (step 2). This view is elliptical in shape. At this point add 3/16-inch-wide notched tabs completely around the development except for the ends. One end will have a regular tab and the other end will be blank. Notched tabs are useful for circular-shape developments. Erase all lettered points and construction lines and "heavy up" all lines. This completes the parallel line development of the truncated cylinder.

Make a copy of the development. Cut out the pattern and fasten it together with glue or tape at the tabbed areas. When finished, check to determine if it resembles the pictorial in Figure 12.11.

12.2.2 The Radial Line Development

The **radial line development** differs from the parallel line development in that all fold lines converge to a single point or radiate from a single point. Therefore, fold lines are not parallel, as they are in parallel line

FIGURE 12.10
Development of a round object.

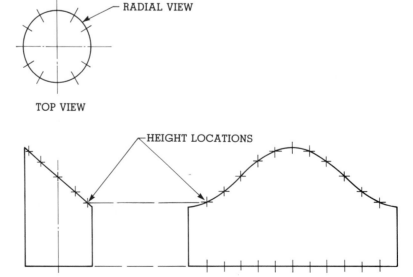

TOP VIEW

FRONT VIEW   DEVELOPMENT

FIGURE 12.11
Procedure for constructing a
development drawing of a
truncated cylinder.
(Continued on page 170.)

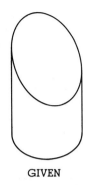

GIVEN

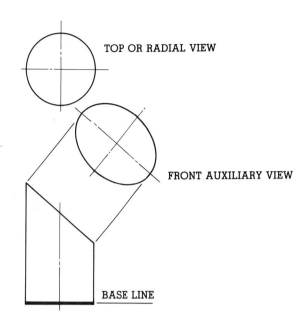

TOP OR RADIAL VIEW

FRONT AUXILIARY VIEW

BASE LINE

FRONT VIEW

STEP 1

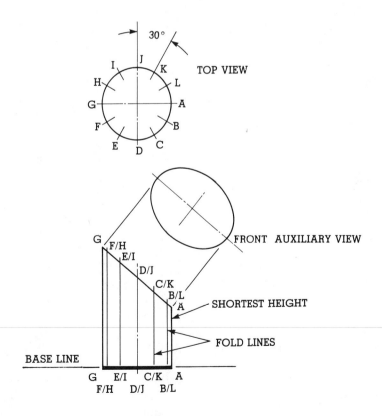

30°

TOP VIEW

I J K L H G A F B E D C

FRONT AUXILIARY VIEW

G F/H E/I D/J C/K B/L A

SHORTEST HEIGHT

FOLD LINES

BASE LINE

G E/I C/K A
F/H D/J B/L

FRONT VIEW

STEP 2

FIGURE 12.11 (continued)

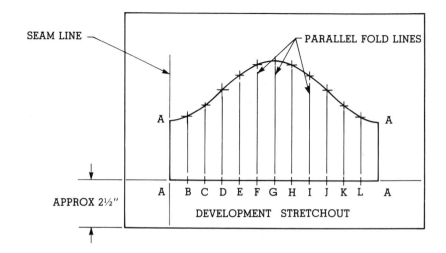

STEP 3

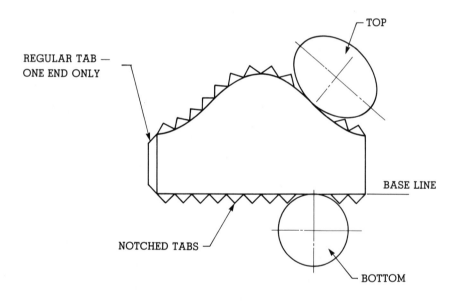

STEP 4

development. The **pyramid** shown in Figure 12.12(a) is a prime example of an object whose development is produced using the radial line method. Note that all four edges and sides meet at a single point.

From the pictorial view in Figure 12.12(a) a two-view orthographic projection is drawn, as illustrated in Figure 12.12(b). The base line and fold lines are shown and identified.

In radial line development the **true length** of each line must be found. In looking at Figure 12.12(c) it is not possible to determine the true lengths of the lines of the pyramid without doing some manipula-

tion of the object. To find the true lengths of the corner lines, rotate the top view of the pyramid until line AOC is on the horizontal center line A'OC'. Projecting point A' downward to the front view allows line A'O to become the true length of line AO. This process is followed to find the true lengths of all the corner lines.

Using the true length A'O as a radius, the flat stretchout development of the pyramid is produced as demonstrated in Figure 12.12(d). The bottom of the pyramid maintains its shape.

The **truncated pyramid** development is con-

FIGURE 12.12
(a) Pyramid; (b) Orthographic projection of pyramid; (c) Determining the true length of a side of the pyramid; (d) Radial line pyramid development.

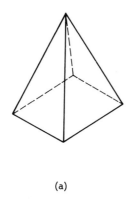

(a)

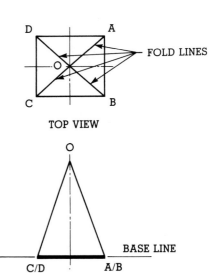

TOP VIEW

FRONT VIEW

(b)

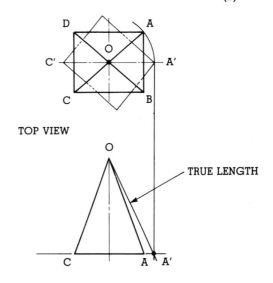

TOP VIEW

FRONT VIEW

(c)

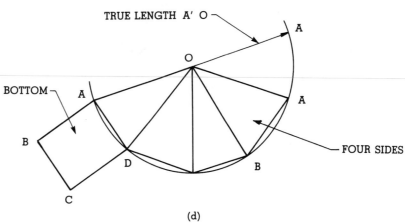

(d)

structed in the same manner as for any other pyramid. The step-by-step procedure is presented below and illustrated in Figure 12.13.

Given: A pictorial view of a truncated pyramid (see Figure 12.13).

Step 1: Draw the front, top, and front auxiliary views. Identify each view as well as the base line and fold lines.

Step 2: To determine the true length of each fold line, first rotate the top view of the object until line

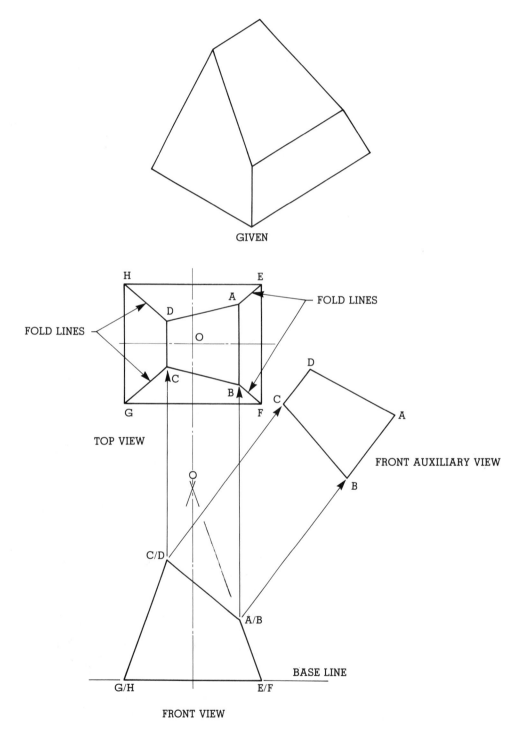

GIVEN

TOP VIEW

FOLD LINES

FOLD LINES

FRONT AUXILIARY VIEW

BASE LINE

FRONT VIEW

STEP 1

FIGURE 12.13
Procedure for constructing a development drawing of a truncated pyramid. (Continued on pages 173 and 174.)

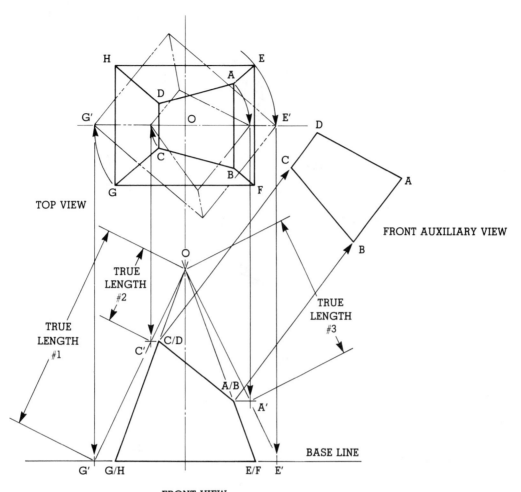

TOP VIEW

FRONT AUXILIARY VIEW

TRUE LENGTH #2

TRUE LENGTH #1

TRUE LENGTH #3

BASE LINE

FRONT VIEW

STEP 2

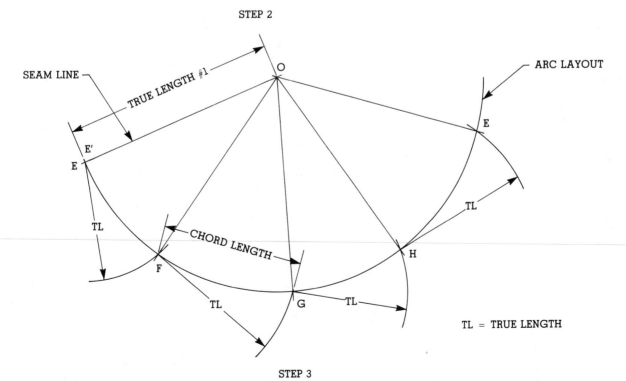

SEAM LINE

TRUE LENGTH #1

ARC LAYOUT

CHORD LENGTH

TL = TRUE LENGTH

STEP 3

FIGURE 12.13 (continued)

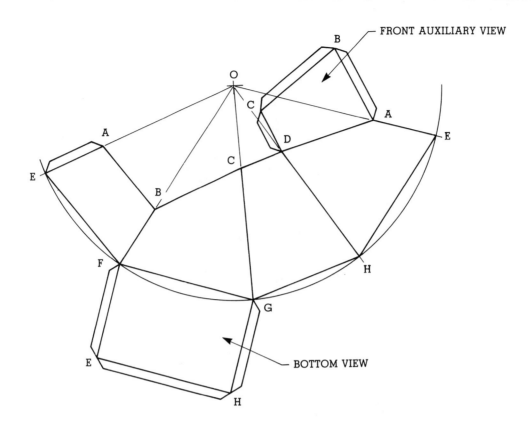

STEP 4

FIGURE 12.13 (continued)

EOG is located on the horizontal center line E'OG'. Project point E' downward to the front view until it crosses the base line. This revolved line from O to E' is now at its true length. In the same manner, project point G' in the top view downward to the front view to obtain the true length OG'. The same procedure is followed to determine true lengths OC' and OA'. Note that four true lengths of corner fold lines have been found: OE', OG', OC', and OA'.

Step 3: This step will provide the sequence for constructing the stretchout development of the pyramid. Making certain that enough space is allowed on the drawing medium, strike arc OE', which is a true length. Using the top view in step 2 as a guide, swing cord length distances from fold line to fold line (E to F, F to G, G to H, H to E). With a light line, connect fold lines OE', OF, OG, OH, and OE' again (because OE' is the seam line).

Step 4: Using the true lengths found in step 3 for reference, locate points A, B, C, D, and A again, along with their respective true length fold lines AE, BF, CG, DH, and seam line AE. Add both the front auxiliary and bottom views, transferring them from step 1. Add 3/16-inch-wide tabs, mak-

ing certain that there are seven tabs and seven blank spaces. When satisfied that the development is complete, erase all construction lines and lettered points and "heavy up" all lines. This completes the radial line development of the truncated pyramid.

Make a white print or a copy of the truncated pyramid development drawing. Cut out the pattern. Fold at the fold lines as well as at the tabbed areas. Fasten the tabbed areas to the body of the pyramid using glue or tape until the assembled truncated pyramid is identical to the pictorial in Figure 12.13.

Radial line development is also needed for producing a stretchout pattern for a **regular cone**. This type of development may be considered as a series of adjacent triangles beginning at the central point, as demonstrated in Figure 12.14(a). When fully developed, the cone stretchout should appear as in Figure 12.14(b).

The development of the **truncated cone** is similar to that of the regular cone except that additional layout work is required because of the truncation. Proceed in accordance with the following sequence

FIGURE 12.14
(a) Adjacent triangles forming
a cone-shape object; (b) Cone
development.

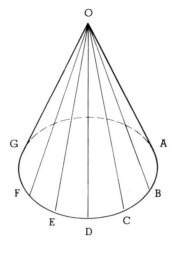

(a)

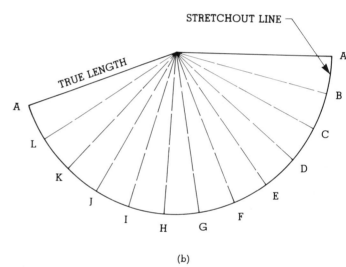

(b)

of steps, referring to Figure 12.15, for constructing a layout of the truncated cone.

Given: A pictorial of a truncated cone (see Figure 12.15).

Step 1: Draw the front and top views. Identify each view as well as the base line. Divide the top view into 12 equal parts. Label the intersection points A through L beginning at the right end of the horizontal centerline and moving clockwise. These divisions represent plotting points for the development of the stretchout and of the inclined (truncated) surface. Project lines downward from the lettered intersections to the front view and label the intersections of these lines with the base line. From these points on the base line, project a set of lines upward so that they converge at **apex** O in the front view. Label these points of intersection at the point of truncation (slant) as A', B'/L', C'/K', D'/J', E'/I', F'/H', and G' as shown. Extend vertical lines upward from these points until they intersect the 12 radial lines that emanate from center point O in

the top view. Label the intersected points in the top view as A', B', C', D', E', F', G', H', I', J', K', and L' as illustrated. Using an irregular curve or template, connect points A' through L' with a light line, taking care to make the curve as smooth as possible.

Step 2: To construct the front auxiliary view of the truncated surface, project perpendicular lines from the edge view of the flat surface to any convenient distance from the front view. The projected points will become A' through L' after being transferred with a divider from the top view. Draw seven horizontal lines A' through G' to the left of the truncation in preparation for drawing the stretchout.

Step 3: On a separate drawing medium, swing an arc equivalent to line OG or OA from the front view in step 2. This represents the true length of the distance from the apex to the base of the truncated cone and is the base line for the stretchout. As is true for developing the truncated cylinder, either of two methods may be used to determine the de-

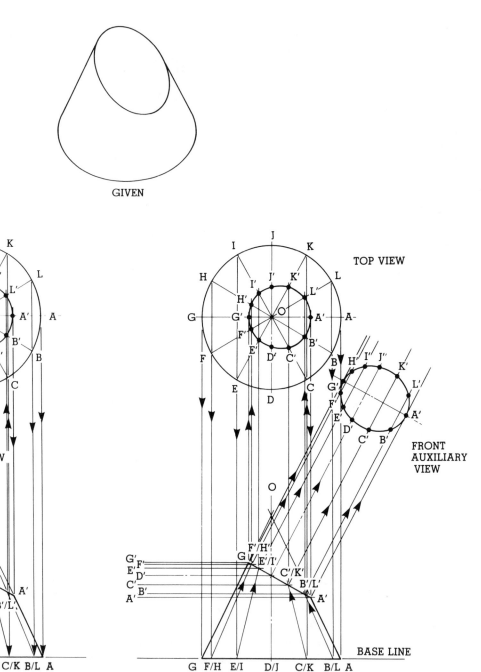

GIVEN

TOP VIEW

FRONT
AUXILIARY
VIEW

TOP VIEW

FRONT VIEW

BASE LINE

STEP 1

BASE LINE

STEP 2

FIGURE 12.15
Procedure for constructing a development drawing of a truncated cone.

velopment of the stretchout: (1) by transferring chordal distances AB, BC, CD, and so on, from the top view and laying them out on the base line; or (2) by the calculation method using the formula for the circumference of a circle (C = 3.14 × di-ameter). Either method is acceptable depending on the accuracy desired. The true lengths for the truncation are located where the seven horizontal lines A' through G', in step 2, intersect true length line OG in the front view of the cone. With dividers,

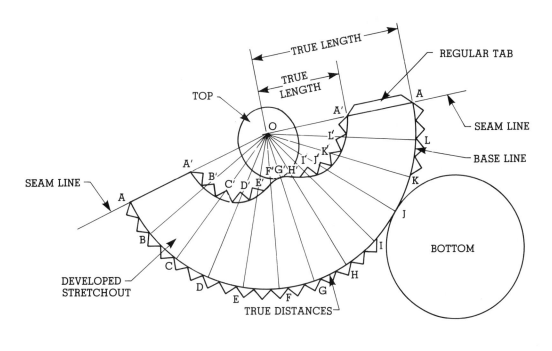

STEP 3

FIGURE 12.15 (continued)

transfer true length OA' to the construction in step 3, as illustrated. Continue the process for OB', OC', OD', OE', and OG' in that order. With an irregular curve, connect the plotted points. Using heavy-weight lines, complete the contour of the development. Add the top, the bottom, and 3/16-inch-wide notched tabs as shown. Also include the straight tab at the seam end. This completes the development of the truncated cone using radial line development.

Make a printed copy of the truncated cone development drawing. Cut out the pattern. Fasten the tabbed areas to the top, bottom, and seam areas of the cone with tape or glue to complete the assembly. Compare the finished model with the pictorial view in Figure 12.15.

12.2.3 The Triangulation Development

The term **triangulation development** is derived from the word **triangle**. It is a method of dividing a surface into a number of triangles and then transferring the true sizes and forms of the triangles into a development drawing. An example of an object for which triangulation development can be used is the **transition piece** (see Figure 12.1). Transition pieces are utilized whenever it is necessary to connect openings of different geometric shapes. The development of a rectangular-to-round transition piece by trian-

gulation development is described below and illustrated in Figure 12.16.

Given: A pictorial of a rectangular-to-round transition piece (see Figure 12.16).

Step 1: Draw the front and top views of the object. Identify each view as well as the base line. Divide the circle in the top view into 12 equal parts. Its origin is O. The intersections are lettered A through L beginning at the top and moving clockwise. The four corners of the rectangle are identified as 1, 2, 3, and 4. Connect the lettered points with the numbered points, producing lines A1, B1, C1, C2, D2, E2, G3, H3, I3, I4, J4, and K4. These divisions represent the triangular-shape sections which will provide the object with a type of curvature.

Step 2: At this point a **true-length diagram** needs to be produced so that the development will be as accurate as possible. A true-length diagram is a means of determining required projections more rapidly than by other means, and is a combination of two views: the top view and the front view. Lay out the true-length diagram by drawing vertical line MN equal to the height of the transition section and horizontal line NP of any convenient length. Using dividers, transfer lines L1, A1, B1, and C1 from the top view to the true-length diagram line NP, and draw lines from these points to point M. The lines LM, AM, BM, and CM in the true-length diagram represent lines L1, A1, B1, and C1,

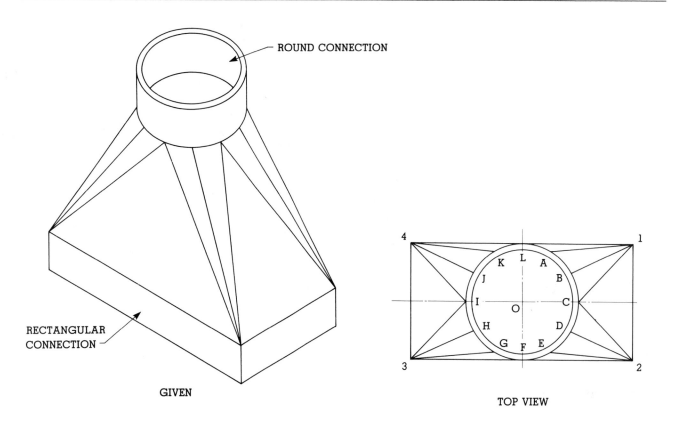

ROUND CONNECTION

RECTANGULAR
CONNECTION

GIVEN

TOP VIEW

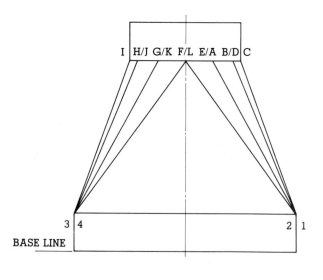

FRONT VIEW

STEP 1

FIGURE 12.16
Procedure for constructing a development drawing of a rectangular-to-round transition piece. (Continued on pages 179 and 180.)

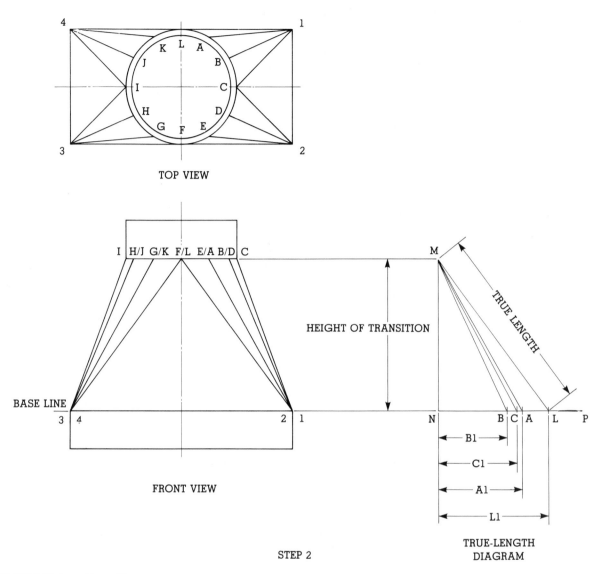

TOP VIEW

FRONT VIEW

HEIGHT OF TRANSITION

BASE LINE

STEP 2

TRUE-LENGTH DIAGRAM

FIGURE 12.16 (continued)

respectively. There is no need to determine additional true lengths in the top view because the other three quadrants of view are duplicates of the true lengths already drawn.

Step 3: For development of the stretchout, use a thick, heavy-weight paper. Allowing sufficient space for the complete development including the round top and straight section at the base, draw vertical line L1 which is equal in length to LM in the true-length diagram. Strike an arc equal to distance LA in the top view and transfer it to the stretchout development. Intersect point A on the stretchout with line A1 which is equivalent to AM in the true-length diagram. Complete the development by adding the remaining intersections and lines in a similar manner, including the round top piece and the base.

The triangulation development of the rectangular-to-round transition piece is now complete. Tabs for fastening the development together may be added, if desired. Cut out the pattern using a scissors or a sharp cutting tool. Score fold lines lightly so that the pattern can be easily formed into its final shape. Check to make certain that the final form is similar to that shown in the pictorial view in Figure 12.16.

12.3 SUGGESTIONS FOR PRODUCING THE DEVELOPMENT DRAWING

The construction of the development drawing, from which a template or pattern is produced, requires much thought, visualization of lines and surfaces, as

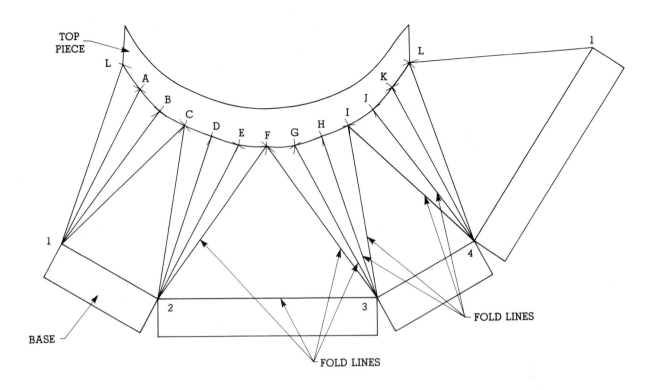

STEP 3

FIGURE 12.16 (continued)

well as the ability to draw and transfer lines and points accurately. Several important guidelines and suggestions for producing the development drawing are listed below.

1. Examine the object carefully to determine what type of development drawing is required—a parallel line development, a radial line development, or a triangulation development.

2. Draw the necessary orthographically projected views, including auxiliary views.

3. Identify each view as well as the fold lines, seam line, base line, and other important points with letters (A, B, C, D) or numbers (1, 2, 3, 4), or both.

4. Try to visualize the object in its open, flat plane—that is, in the stretchout developed form. Make a freehand sketch if necessary.

5. Determine if the development of true lengths of lines is required or if a true-length diagram will be necessary.

6. Begin the actual development layout on a separate drawing medium. Use dividers and/or a compass for transferring distances and striking arcs. Follow an established sequence of steps for producing the pattern so as not to be confused if it is a complex layout.

7. After the layout is complete, decide if tabs are required for completing the model of the object. If tabs are required, determine whether regular or notched tabs are best suited for the object.

8. Cut out and assemble the developed pattern, and see if it meets the intended need.

12.4 SUMMARY

The ability to produce a quality development drawing is an important, necessary skill for the drafter. In this section, three basic types of development drawings were discussed: the parallel line development, the radial line development, and the triangulation development. The purpose of the development drawing was established, and examples of items for which development drawings are required were presented.

It was explained that objects with single-curved surfaces that can be unfolded or rolled out without distortion are considered to be "developable." Several examples of developable objects were illustrated.

The text material covered the development of simple rectangular objects, truncated prisms, truncated cylinders, regular and truncated pyramids, regular and truncated cones, and a rectangular-to-round transition piece. In addition, several technical terms peculiar to development drawings were presented.

Finally, several suggestions were offered as aids to the drafter in producing development drawings. Practice exercises for this section are presented on pages 12A through 12H in the workbook.

12.5 **REVIEW EXERCISES**

1. What is the purpose of the development drawing?

2. Give four examples of objects that require development drawings.

3. List the three major types of developments used to produce a pattern or template of an object.

4. Sketch a truncated cylinder in the space provided. What kind of development is used to develop a pattern for this cylinder?

5. Give an example of an object whose development drawing is produced by the radial line method.

6. In parallel line development, at what angle must the base line be to fold lines?

7. In what way does radial line development differ from parallel line development?

8. Explain the difference between a regular cylinder and a truncated cylinder.

9. What is the purpose of tabs?

10. What are the names of two types of tabs?

11. Give an example of an object that requires triangulation development.

12. Define "transition piece."

13. What is a true-length diagram?

14. What kind of development drawing is required to construct the stretchout of a truncated prism?

15. List six suggestions and guidelines for producing a development drawing.

    a.

    b.

    c.

    d.

    e.

    f.

$+ \ xF$ bF   bR $+ \ xR$

dF   dR

cF   cR

aF   aR

TRUE LENGTH

F | R

Section 13

Descriptive Geometry

LEARNER OUTCOMES

The student will be able to:

- List the basic principles of descriptive geometry

- Identify technical terms used in descriptive geometry

- Provide graphical solutions for seven basic types of problems in descriptive geometry

- Discuss several suggestions for successfully solving problems in descriptive geometry

- Demonstrate learning through the successful completion of review and practice exercises

13.1 PURPOSE

For a drafter or designer, the purpose of learning the principles and standard practices of descriptive geometry is to become more effective in solving problems involving spatial relationships and graphical analysis. Adoption of these practices and methods of construction, and familiarity with the common terms associated with them, will greatly facilitate the drafter in interpreting the technology of drafting. A good basic grounding in this topic will be valuable to the computer-aided drafting (CAD) operator in the development and visualization of three-dimensional objects, complex part features, and rotated parts.

Descriptive geometry may be defined as the practice of projecting three-dimensional objects onto a two-dimensional medium so as to solve problems involving lines, angles, and shapes by the use of drafting technology.

13.2 BASIC PRINCIPLES

Sections 11 and 12 (The Auxiliary View and The Development Drawing) have provided the basic concepts and experiences which should be sufficient to permit the learner to advance into the topic of descriptive geometry. As is true for most new learning experiences, basic principles should be understood and internalized. The basic principles identified below will assist the learner in the solution of problems in descriptive geometry.

1. In orthographic (third-angle) projection (see Section 6, The Multiview Drawing), the **plane** is always between the object and the observer, with the object changing position to accommodate the observer. In descriptive geometry, the observer changes position, not the object.
2. For the various views required, **projection planes** are always rotated into the plane of the drawing through a 90-degree angle. Rotation of the object takes place about lines which may be seen as hinged between adjacent projection planes. These lines which fold or rotate are called **reference lines**, as shown in Figure 13.1. Reference lines are to be used as temporary construction lines and should not appear on the finished drawing.
3. For separate observation of each view, reference lines represent edge views of other projection planes.
4. The **lines of sight** from the object to each projection plane are parallel with each other and perpendicular to the projection plane.
5. Perpendicular distances from the projection plane to the object are **equal distances** in all views perpendicular to that plane.

13.3 TERMINOLOGY USED IN DESCRIPTIVE GEOMETRY

There are several important terms related to descriptive geometry with which the learner should become familiar. These terms are defined below.

- **Notation** A systematic means of identifying points and views on a drawing. For example, the learner may wish to use the following guidelines in the solution of graphical problems.
  - Use lower-case letters (a, b, c, etc.) for all points.
  - Use upper-case letters for all views. (Example: F = front view; T = top view; L = left side view; R = right side view; A = auxiliary view(s); B thru Z = all other views.)
  - Two or more points may be combined to locate a line.
  - Upper- and lower-case letters may be combined to identify a point on a specific view.
- **Plane** Normally refers to a surface, such as a flat **plane surface**. It is the place, anywhere on a surface, where two points connect to form a straight line. A flat plane is a surface which is not curved or warped.
- **True shape** The true and actual shape of a surface.
- **True length** The true and actual length of a line.
- **Edge view** A view of a plane wherein all lines of the plane appear to coincide in a single line.
- **Line of sight** An imaginary straight line from the eye of the observer to a point on the object being observed.
- **Point view** A view of a line wherein the line of sight is parallel with the line, resulting in a single point.
- **Normal view** A view of a plane wherein the line of sight is perpendicular to the plane.

13.4 DETERMINING THE TRUE LENGTH OF A LINE

The true length of a line has a projection which is parallel with a principal plane of projection. Thus a line parallel with a plane of projection is shown at its true length in that projection. When two or more views of a line are given, none of which is a true view, an additional view is required to show the true length. This is accomplished by drawing a **reference line** parallel with one of the given views and projecting the line onto a plane, as outlined in the following solution and illustrated in Figure 13.2.

Given: The front, top, and right side views of line ab with reference lines drawn at 90 degrees between views, as depicted in Figure 13.2.

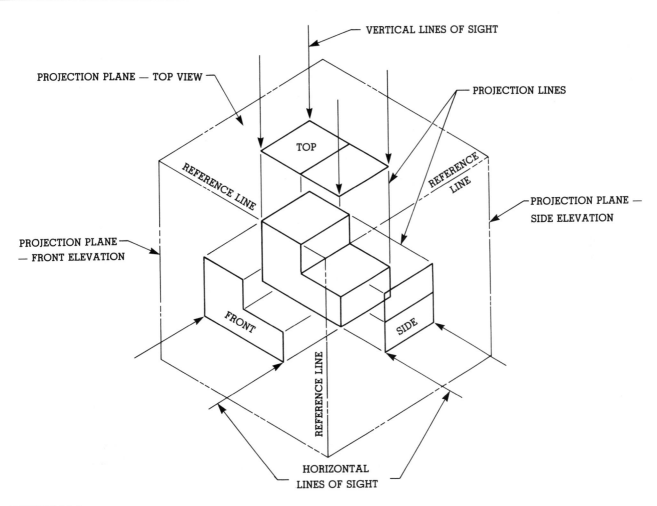

FIGURE 13.1
Pictorial illustration of basic principles of descriptive geometry.

Step 1: Draw a reference line parallel with line ab on the right side view, at any reasonable distance, and label it as shown. Project light lines at 90 degrees to the reference line from points a and b as shown.

Step 2: Using dividers, determine distances x and y from the vertical reference line to the front view and transfer them to form an auxiliary view. The length of the resultant line is the true length of line ab. Identify all points and lines.

13.5 DETERMINING THE POINT VIEW OF A LINE

In many problems, determining the point view of a line is the next step after finding its true length. Point views of lines are very useful in the solution of problems related to spatial geometry.

Given: The front, left side, and auxiliary views of line ab (see Figure 13.3). These views are separated by reference lines L/F and F/A. The auxiliary view

shows the true length of line ab, which must be determined before proceeding to determine the point view.

Step 1: At a reasonable distance from one end of the true length of line ab, draw reference line A/B perpendicular to line ab. Extend the true length with a projected line through the perpendicular line, as shown.

Step 2: Using dividers, establish a point on the extension of the true length of line ab that is distance x (the distance from reference line F/A to the front view of line ab) from the intersection of the extended true length line and perpendicular bisector A/B. This point, abB, is the point view of line ab. Label all points and lines.

13.6 DETERMINING THE EDGE VIEW OF A PLANE

A plane surface becomes an edge, or a straight line, on the view showing the true length in that plane as a

FIGURE 13.2
Procedure for determining the
true length of a line.

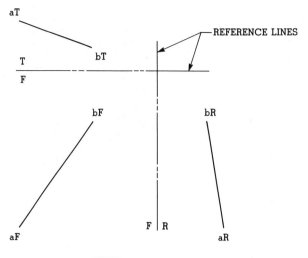

GIVEN

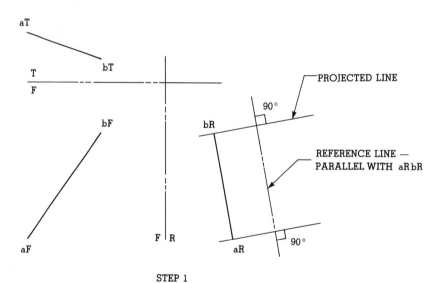

STEP 1

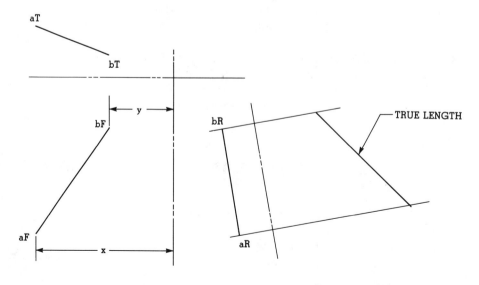

STEP 2

FIGURE 13.3
Procedure for determining the
point view of a line.

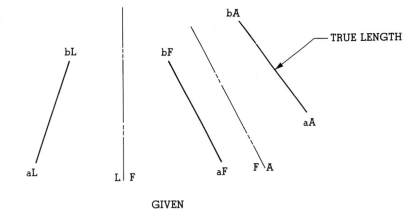

TRUE LENGTH

GIVEN

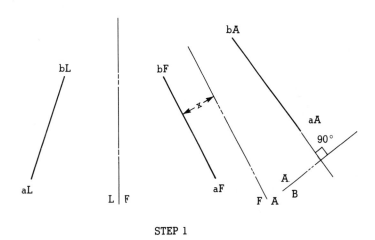

STEP 1

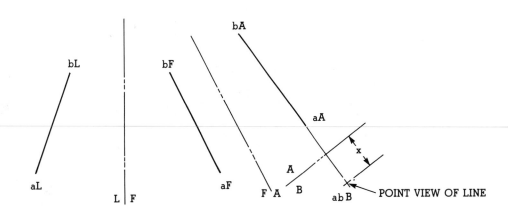

POINT VIEW OF LINE

STEP 2

point view. In order to develop an edge view, the line must first be drawn in the plane such that the line appears in its true length in an adjacent view.

Given: The front and top views of plane abc and a horizontal reference line between views (see Figure 13.4).

Step 1: Draw line dc in the top view of the plane. This

line is drawn parallel with reference line T/F. Drop light vertical lines from d and c in the top view until they touch the front view at d and c, respectively. Line dc appears at its true length in the front view.

Step 2: Construct reference line F/A perpendicular to the true length of dc. Using dividers, establish points aA, dA/cA, and bA at distances from reference line F/A equal to the distances of the corre-

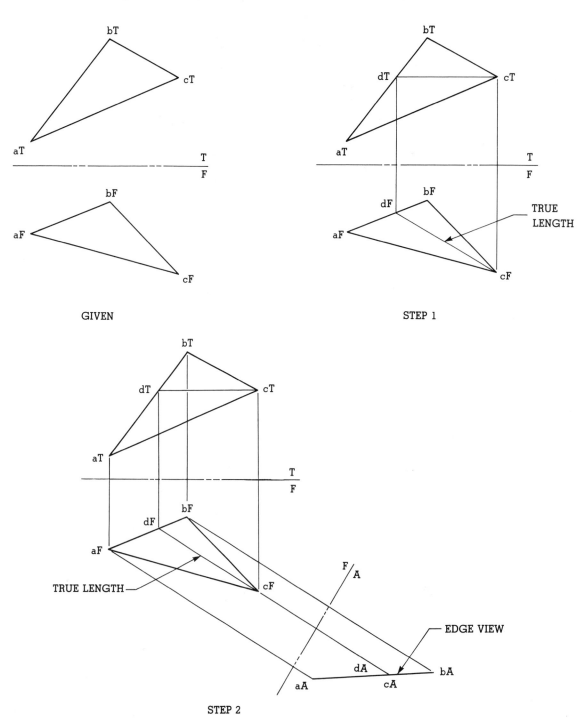

GIVEN

STEP 1

STEP 2

FIGURE 13.4
Procedure for determining the edge view of a plane.

FIGURE 13.5
Procedure for determining the
true size and shape of a plane.

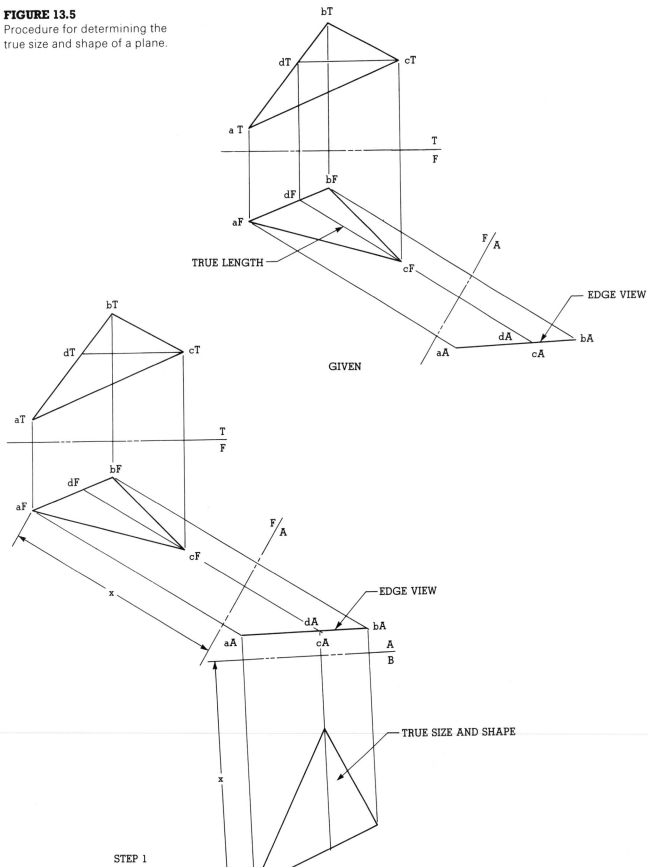

TRUE LENGTH

EDGE VIEW

GIVEN

EDGE VIEW

TRUE SIZE AND SHAPE

STEP 1

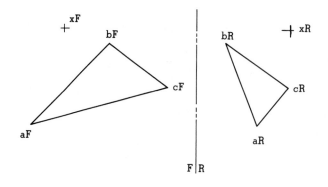

GIVEN

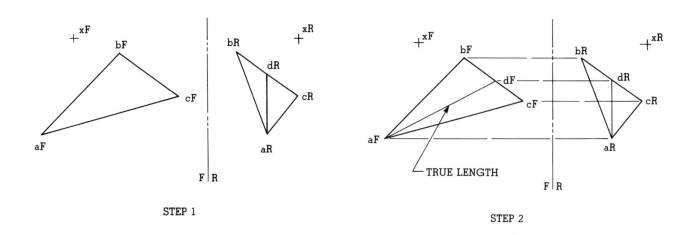

STEP 1

STEP 2

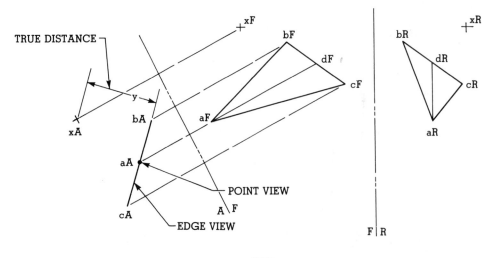

STEP 3

FIGURE 13.6

Procedure for determining the true distance between a plane surface and a point in space.

sponding points in the top view from reference line T/F. Joining these points to form a straight line will result in an edge view of plane abc.

13.7 DETERMINING THE TRUE SIZE AND SHAPE OF A PLANE

In order to determine the true size and shape of a plane, it is first necessary to determine a true length, a point view, and an edge view. In this problem we continue our determination of the edge view and build on what was previously done to produce a view showing the true size and shape of plane abc.

Given: Plane abc with its edge view and reference lines between views, as illustrated in Figure 13.5.

Step 1: Draw reference line A/B parallel with the edge view of the plane. Using dividers, project points a, b, and c using distance x as a guide. The result shows the true size and shape of plane abc.

13.8 DETERMINING THE TRUE DISTANCE BETWEEN A PLANE SURFACE AND A POINT IN SPACE

Determining the true distance between a given plane and a point in space requires a view that shows both the edge view of the plane and the point in space. The solution is shown in Figure 13.6.

Given: The front and right side views of plane abc and point x, with a vertical reference line (F/R) between views.

Step 1: In the right side view, draw a line parallel with reference line F/R beginning at aR. Label this line as aRdR.

Step 2: Develop the true length of line aRdR in the front view. Identify it as aFdF.

Step 3: Extend the true length line aFdF into the auxiliary view. Also project into this view the other points of the plane surface and point x, making certain that they are projected perpendicular to reference line A/F. Produce the edge view of the plane by the method outlined in Subsection 13.6. The true distance between the plane surface and the point in space may be obtained by measuring the perpendicular distance y.

13.9 DETERMINING THE TRUE ANGLE BETWEEN TWO PLANE SURFACES

The true angle between two plane surfaces will be observed in a view which shows the edge views of

both planes. Because the intersection line is common to both planes, the end view of the intersection line will provide the edge views of both planes.

Given: Front, side, and top views of plane surfaces abc and abd, which intersect at edge ab. The three views are separated by perpendicular reference lines (see Figure 13.7).

Step 1: Using the knowledge obtained from previous problems, determine the true length of the edge of intersection ab. Project all points into an auxiliary view, A.

Step 2: Find the point view of the true length of the edge of intersection ab. Project all points into a secondary auxiliary view, B. The resultant true angle between the two planes can be measured to produce the solution to the problem.

13.10 DETERMINING THE INTERSECTION OF TWO PLANES USING THE LINE PROJECTION METHOD

When the intersection of two planes is found by determining the edge view of one of the planes, it is called the intersection by auxiliary method. A simpler method, which is outlined below, is the **line projection** or **piercing point method**. This method is frequently used to solve problems of developments involving the intervention of plane surfaces.

Given: Top and front views of plane surfaces abc and def, as illustrated in Figure 13.8.

Step 1: Locate points x and y in the top view on edge ac. Project points x and y downward to the front view, and draw a light line xy in the front view. The point where xy intersects the edge line ac is called the **piercing point**. Follow the same process to locate the piercing point for edge line ab.

Step 2: After finding all the piercing points in both the front and top views, complete the views as shown. The solution clearly identifies the delineation of both views and how and where the planes intersect.

13.11 SUGGESTIONS FOR SOLVING PROBLEMS IN DESCRIPTIVE GEOMETRY

The tasks of drafters and designers include solving problems involving relationships. Among them are geometric questions about points, lines, planes, and intersections. Solutions to these problems can be obtained mathematically or, in our case, by using the technology of drafting. They also can be solved using conventional drafting practices or by the use of

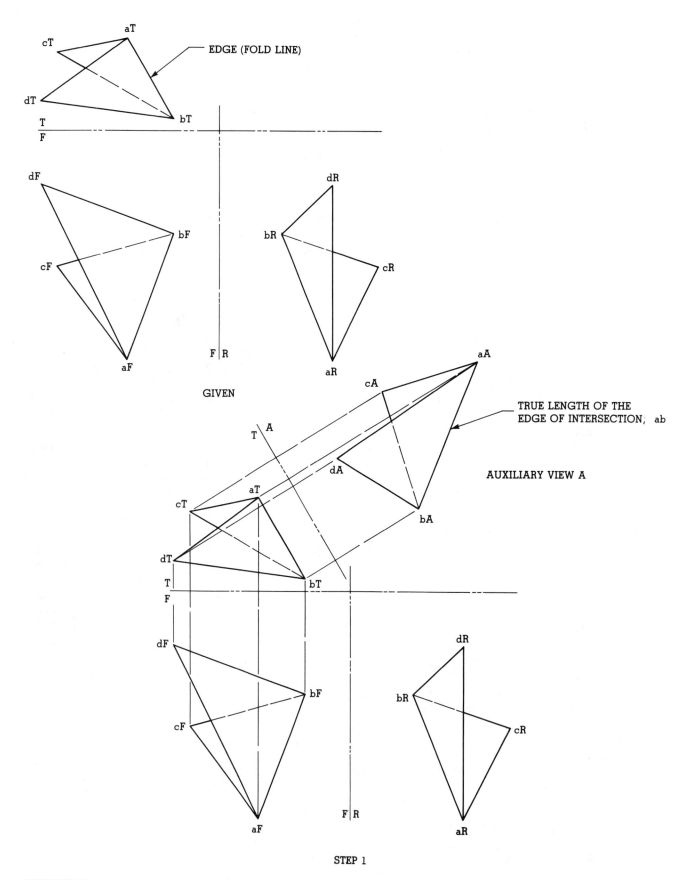

EDGE (FOLD LINE)

GIVEN

TRUE LENGTH OF THE
EDGE OF INTERSECTION; ab

AUXILIARY VIEW A

STEP 1

FIGURE 13.7
Procedure for determining the true angle between two plane surfaces.

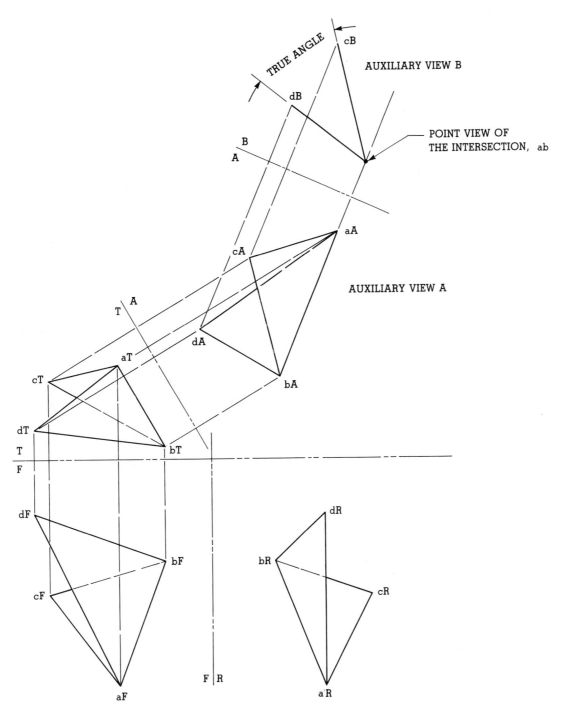

STEP 2

FIGURE 13.7 (continued)

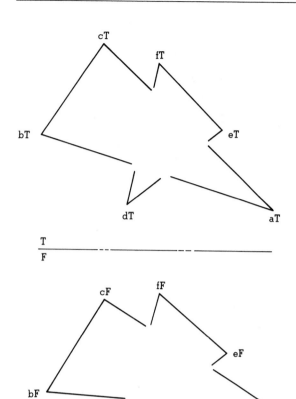

GIVEN

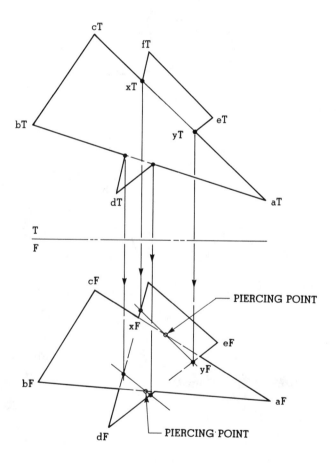

STEP 1

FIGURE 13.8
Procedure for determining the intersection of two planes using the line projection method.

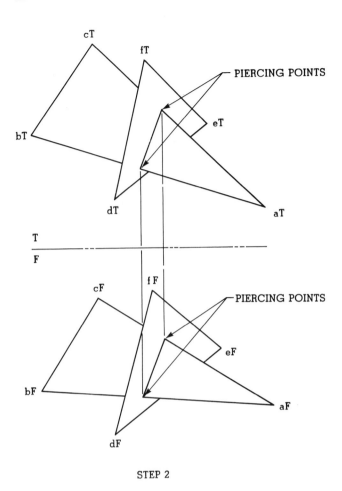

STEP 2

FIGURE 13.8 (continued)

computer-aided drafting (CAD). The following suggestions will aid the learner in the solution of basic problems in descriptive geometry.

1. Select the appropriate medium on which the problem will be drawn and the solution completed.

2. Carefully review the problem. Be knowledgeable about the technical terms used. Identify the given information. An important first step in problem solution is determining what is given.

3. Establish a system of notation to be used so that one can follow subsequent steps easily.

4. Most solutions will involve the following, in the order given:
   a. Determining the true length of a line
   b. Determining the point view of a line
   c. Determining the edge view of a plane
   d. Determining the true size and shape of a plane

5. Do all that is required at each sequence before continuing to the next step. Identify all points, lines, and planes after each step.

6. Use dividers to transfer measurements. Make certain that each measurement is obtained from the correct view and from the appropriate reference line.

7. When satisfied with the solution to the problem, darken all lines that represent the solution. Erase unwanted lines.

13.12 **SUMMARY**

Solving problems involving descriptive geometry is a useful skill for those working with manual drafting equipment and traditional practices of drafting technology as well as for those engaged in computer-aided drafting (CAD). A basic knowledge of orthographic projection as well as familiarity with auxiliary views and development drawings are helpful prerequisites in the solution of geometric problems covered in this section.

Basic principles and technical terminology were detailed and several problems were identified and solved using a clear, concise method of presentation.

Finally, suggestions for aiding the drafter in the solution of problems involving descriptive geometry were presented. Practice exercises for this section are presented on pages 13A through 13H in the workbook.

13.13 REVIEW EXERCISES

1. What is the purpose of learning to solve problems in descriptive geometry?

2. State the basic principles utilized in descriptive geometry.

3. Define the following technical terms:

Plane

Edge view

Light of sight

Point view

4. Explain the steps involved in determining the true length of a line.

5. What is a reference line and what does it represent?

6. What care should be taken when transferring measurements in the solution of a problem?

7. Explain the steps involved in determining the intersection of two planes.

8. What are notations? How are they used?

9. What is the basic difference between orthographic projection and descriptive geometry, in the eye of the observer?

10. List four helpful hints in solving problems of descriptive geometry.

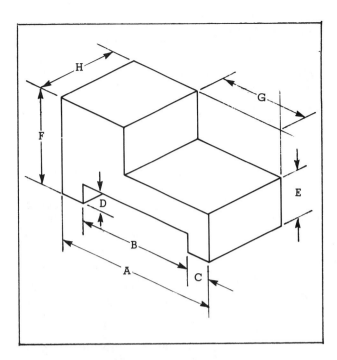

Section **14**

The Pictorial Drawing

LEARNER OUTCOMES

The student will be able to:

- Describe the characteristics of a pictorial drawing

- List three types of pictorials

- Identify the fundamental differences and similarities among isometric, oblique, and perspective drawings

- Construct pictorial drawings using a step-by-step process

- Explain the difference between isometric and non-isometric lines

- Identify three types of oblique drawings

- Describe the angular, two-vanishing-point perspective drawing

- Define technical terms used in the construction of a perspective drawing

- Demonstrate learning through the successful completion of review and practice exercises

14.1 PURPOSE

The **pictorial drawing** is very descriptive and graphic because it portrays what the eye sees when viewing an object. This kind of drawing presents an effective method of communicating a concept and shows all three dimensions of an object in a single view. It is especially helpful for the nontechnical or inexperienced person in reading and understanding what an object looks like or how parts fit together. Pictorial drawings are used by engineers, designers, and drafters to visualize designs better; by production and assembly personnel to determine the sequence of assembly operations; by quality control personnel to facilitate inspection procedures; and by field technicians to facilitate the installation and replacement of parts.

This section will cover the three most commonly used pictorial drawings: (1) the isometric projection, (2) the oblique projection, and (3) the perspective projection. The fundamental differences among the three are illustrated in Figure 14.1.

14.2 THE ISOMETRIC DRAWING

The word **isometric** means "equal measure" and is used to describe this type of pictorial drawing. The isometric is a drawing shown on a single plane or flat surface but in reality is a three-dimensional drawing because height, width, and depth are all pictorially described. In its true form, an isometric projection should be drawn at approximately 80% of its actual size, but in normal practice it is drawn full size. For our purpose the isometric drawing will be drawn full size. Isometric templates (see general discussion of templates in Section 2, Use of Graphics Equipment and Tools) are available to aid the drafter in the production of isometric drawings.

14.2.1 The Isometric Axis

The isometric drawing is produced by first drawing three **axes**: one vertical axis, and two axes drawn to the left and right of the vertical, each at a 30-degree angle from a horizontal line, as depicted in Figure 14.2.

Another way of describing the isometric axes is to say that they are shown as three principal edges of an object that are drawn at equal 120-degree angles. A 30/60-degree triangle or a drafting machine is used to determine the isometric axes, as shown in Figure 14.3.

14.2.2 The Isometric Line

Lines that are drawn parallel to the three axes of an isometric drawing are called **isometric lines**. The ac-

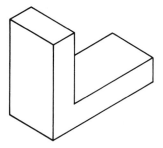

ISOMETRIC PROJECTION

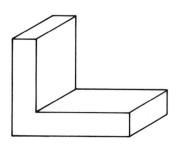

OBLIQUE PROJECTION

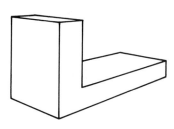

PERSPECTIVE PROJECTION

FIGURE 14.1
Isometric, oblique, and perspective projections.

tual width, height, and depth measurements of an object are taken directly from the orthographic views shown in Figure 14.4 and transferred to the isometric drawing to become corresponding isometric lines. True measurements can be made along isometric lines or lines that are parallel with them.

On an isometric drawing, hidden lines are not normally shown, unless they are critical for figure description.

14.2.3 The Nonisometric Line

The **nonisometric line** is one that is not parallel with any isometric axis. Objects that have sloping or slanted surfaces have lines that are nonisometric in nature. Lines of this type do not appear in their true lengths and, therefore, cannot be measured directly,

FIGURE 14.2
Three isometric axes.

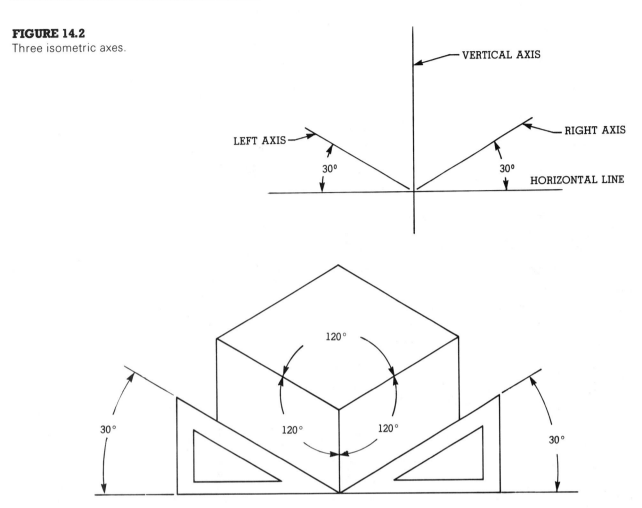

FIGURE 14.3
Projecting the isometric axes.

as can isometric lines. A preferred method for drawing nonisometric lines or surfaces is called the **box method**, which is demonstrated by the step-by-step process outlined below and illustrated in Figure 14.5.

Given: Two orthographic views of object A, with its slanted surface identified by points 1, 2, 3, and 4 (see Figure 14.5).

Step 1: Using light-weight construction lines, draw object A in its basic isometric form.

Step 2: Using dividers or a scale, transfer directly the measurements X and Y from the given orthographic view of the object, and locate points 1, 2, 3, and 4 on the basic isometric form.

Step 3: Complete the isometric drawing of object A by using the correct line weight for visible lines. The finished drawing shows both isometric and nonisometric lines.

14.2.4   The Irregular Isometric Curve

The **irregular isometric curve** is also considered a nonisometric line. This type of curve can be deter-

mined only by a series of reference lines and plotted points to develop its shape. As is the case for all nonisometric lines, the true shape of the irregular isometric curve must first be drawn in the orthographic view as shown in Figure 14.6, then transferred to the isometric view by direct measurement.

Given: Two orthographic views of an object B, which has an irregular curved surface (see Figure 14.6).

Step 1: Draw several horizontal lines across the object; in our case eight lines are used. The more lines that are used, the smoother the irregular curve should be at completion. The lines are numbered clockwise from 1 to 16 where they intersect the curved surface, as shown.

Step 2: Using light-weight construction lines, draw object B in its basic isometric form.

Step 3: Transfer the horizontal lines from the orthographic drawing to the isometric drawing using the same spacing of lines. Show numbered termination points for all eight lines.

Step 4: Connect the plotted points using an irregular curve or curved template, until the irregular isometric curve is formed.

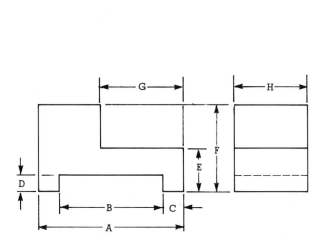

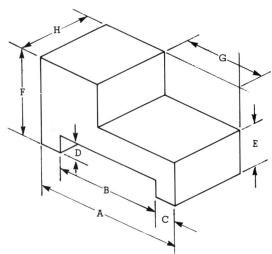

FIGURE 14.4
Transferring measurements to an isometric drawing.

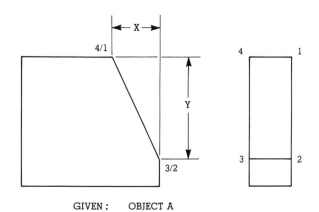

GIVEN : OBJECT A

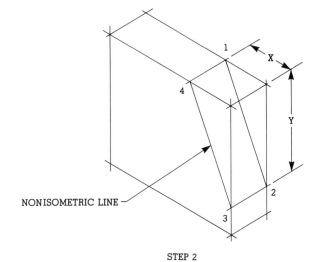

NONISOMETRIC LINE

STEP 2

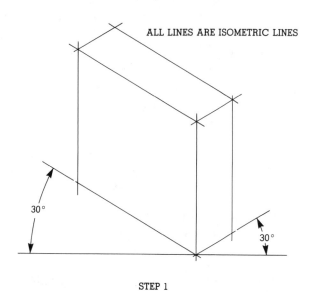

ALL LINES ARE ISOMETRIC LINES

30°

30°

STEP 1

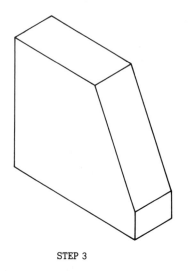

STEP 3

FIGURE 14.5
The box method for constructing nonisometric lines.

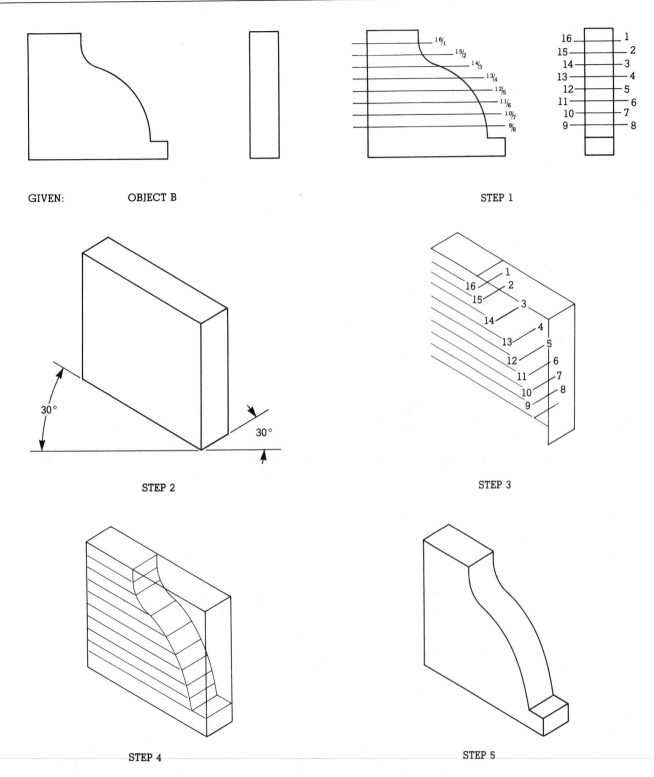

GIVEN: OBJECT B   STEP 1

STEP 2   STEP 3

STEP 4   STEP 5

FIGURE 14.6
Procedure for constructing an irregular nonisometric curve.

Step 5: Complete the isometric drawing of object B by using the correct line weight for visible lines. The finished drawing shows both isometric lines and irregular, nonisometric curved lines.

14.2.5 The Isometric Circle and Arc

On an isometric drawing, a circle takes the shape of an **ellipse** and an arc becomes a **partial ellipse**, as illustrated in Figure 14.7.

FIGURE 14.7
(a) Circle and arc shown in
basic form; (b) Circle and arc
shown in isometric form.

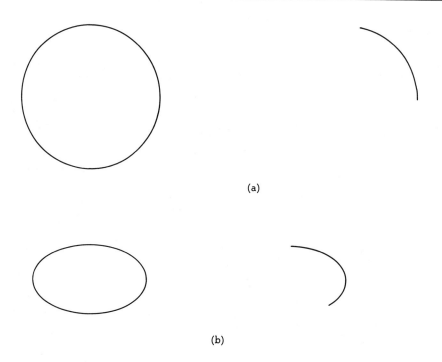

(a)

(b)

Because an ellipse is formed using a vertical line and angled lines at 30 degrees, there are three faces or planes in which an ellipse can be shown. Regardless of the view in which a circle or arc is located, it can be drawn in only one of the three planes, shown in the cube model in Figure 14.8. (Figure 14.8 should be used as a reference for isometric practice exercises for Section 14 in the workbook.)

There are several methods for producing the **isometric circle**. The approach outlined below and illustrated in Figure 14.9 is called the **four-center system**. It is the method most often used, takes less

FIGURE 14.8
Isometric circles and arcs in
various planes.

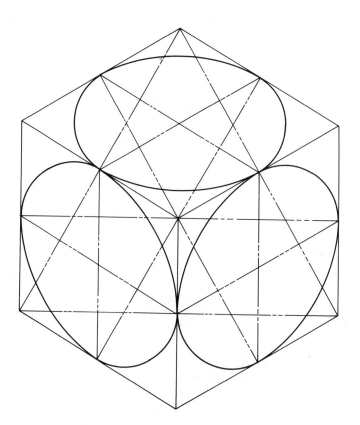

time to construct than other methods, and is sufficiently accurate for most isometric drawing applications. Templates are used when accuracy is not critical.

Given: The basic forms of regular circle A, whose radius is r, and isometric circle B, both with horizontal and vertical center lines whose points are identified as 1, 2, 3, and 4 (see Figure 14.9). *Note*: Points 1 through 4 are tangent points.

Step 1: Draw a horizontal center line. Construct a vertical isometric center line at a 60-degree angle with a triangle or a drafting machine. Using the given radius r from the regular circle, strike radius r on the construction, making certain that the radius intersects the center lines at points 1, 2, 3, and 4.

Step 2: Construct the basic isometric form. Draw four lines at 90 degrees, one from each tangent point, as shown. Note that these four lines intersect at two places. Identify these intersections as C and D.

Step 3: Strike minor radii from points C and D and major radii from points E and F, using light-weight

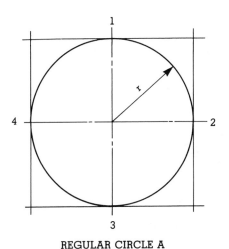

REGULAR CIRCLE A

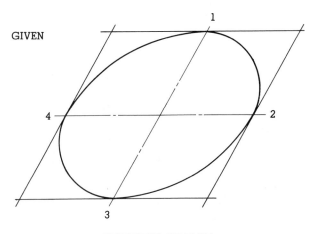

GIVEN

ISOMETRIC CIRCLE B

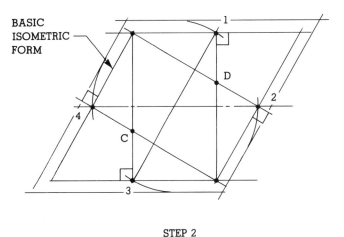

STEP 2

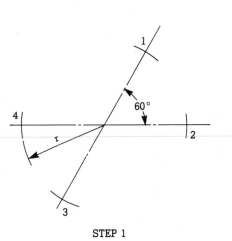

STEP 1

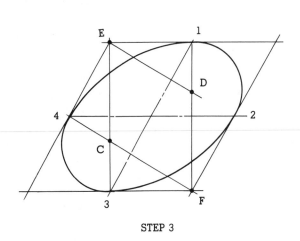

STEP 3

FIGURE 14.9
The four-center system for constructing the isometric circle.

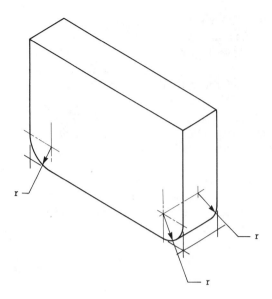

FIGURE 14.10
Drawing the isometric arc.

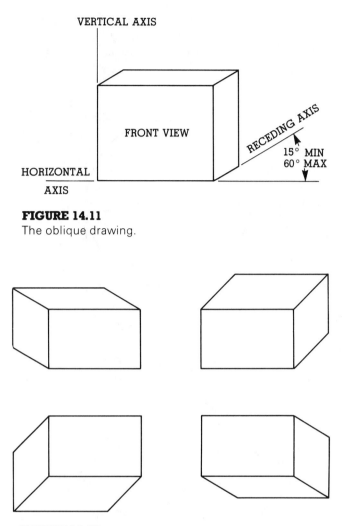

FIGURE 14.11
The oblique drawing.

FIGURE 14.12
Typical receding axis positions.

lines. Each of the four radii equals one-fourth of the isometric circle. When satisfied that all tangent points meet, "heavy up" all arcs, using the appropriate line weight. The isometric circle is now complete.

The **isometric arc** can be drawn with the same four-center system used for producing the isometric circle. However, it is not necessary to draw the full construction. Only the radius for drawing the arc is required, as illustrated in Figure 14.10.

14.3 THE OBLIQUE DRAWING

There are three types of **oblique drawings**: the **cabinet type**, where the receding axis is drawn at an angle of 30 to 60 degrees; the **general type**, which is similar to the cabinet type except that its receding distances are drawn at any angle; and the **cavalier type**. The most commonly produced oblique drawing, and the one which will be discussed in this section, is the cabinet type.

All oblique drawings consist of three axes: the **vertical**, the **horizontal**, and the **receding**. The receding axis may be drawn at any angle from 15 to 60 degrees from the horizontal axis, as shown in Figure 14.11. The front view is the most significant view and is drawn to its true scale and shape.

The receding axis can face to the left, right, above, or below the front view. Figure 14.12 illustrates several positions of the receding axis for oblique projection.

In the cabinet drawing, the receding distances are drawn at half size—that is, are shortened to one-half of their true dimensions—to more closely resemble what the human eye would actually see and to eliminate distortion. Figure 14.13 illustrates the process, outlined below, for producing the cabinet-type oblique drawing.

Given: A three-view orthographic projection of object A with dimensions X, Y, and Z (see Figure 14.13).

Step 1: Draw the true full-scale shape of the front view of object A by measurement and transfer of scaled dimensions.

Step 2: Draw the receding axis at a convenient angle (between 30 and 60 degrees). In our case we will specify 45 degrees. Measure the depth of the object to one-half its actual dimension (Y/2).

Step 3: Complete the cabinet-type oblique drawing by eliminating unwanted extension lines. "Heavy

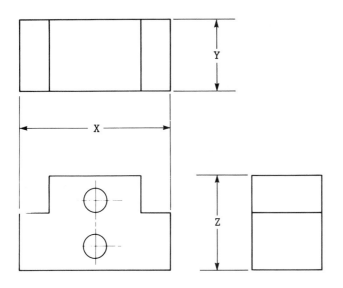

FRONTAL VIEW

GIVEN: OBJECT A

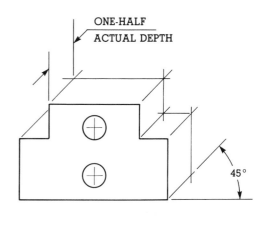

STEP 2

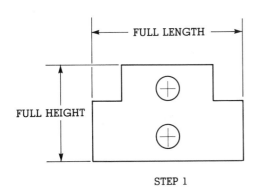

STEP 1

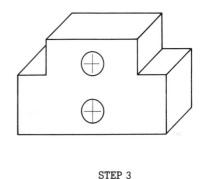

STEP 3

FIGURE 14.13
Procedure for producing a cabinet-type oblique drawing.

up" visible lines using the appropriate line weight. Note that hidden lines are not normally shown on the oblique drawing.

14.3.1 The Oblique Angled Surface

Slanted surfaces and angles on oblique drawings may be drawn using the box method previously discussed for the isometric drawing (see Figure 14.5). Angles which are parallel with the front view are drawn at true size. All other slanted surfaces are determined by locating the ends of the inclined lines, as illustrated in Figure 14.14.

14.3.2 The Oblique Circle and Arc

Whenever possible, surfaces with holes in them should be drawn as the front views in oblique drawings so as to avoid having to draw ellipses for circles. When circles need to be drawn in other than the front view, the four-center ellipse system previously discussed (see Figure 14.9) may be followed. Note that the only difference for the oblique ellipse is that the center of the major radius may be located outside or at some point inside the parallelogram depending on the angle of the receding axis. If the receding axis is at an angle greater than 45 degrees to the horizontal, the major radius center point will be outside the parallelogram. If the angle is less than 45 degrees, the major radius center point will be inside the parallelogram. Figure 14.15 demonstrates these differences.

If available, a template should be used for producing an oblique ellipse, because it will reduce drawing time and provide much better results. When a template is used, an oblique square should first be drawn lightly to locate the circle in the center position. Then the proper size and shape of the ellipse should be selected from the template.

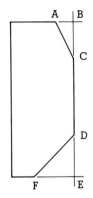

ORTHOGRAPHIC VIEW

(a)

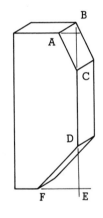

OBLIQUE VIEW

(b)

FIGURE 14.14
Drawing angled surfaces: (a) Orthographic view;
(b) Oblique view.

14.4 THE PERSPECTIVE DRAWING

The **perspective drawing** is the result of a projection technique for representing three-dimensional objects and depth relationships on a flat plane or surface. The perspective drawing differs from the isometric or the oblique drawing because it is the most realistic relative to what the eye really sees. It shows an object as it actually appears instead of showing its true size and shape. It is used extensively by architects to illustrate buildings, interior designs, and landscaping plans. It is also used by technical illustrators to show exploded views and illustrations for technical publications and to prepare graphics for advertising. A good example of a perspective view is what an observer would see when standing on straight railroad tracks on flat terrain and looking down the tracks as far as the eye could see. To the observer it would appear that the two tracks eventually meet at a single point. The point where the observer is standing is called the **station point**. The point on the horizon where the railroad tracks appear to meet is called the **vanishing point**.

There are three kinds of perspective drawings: (1) the **parallel** or **one-vanishing-point** type; (2) the **angular** or **two-vanishing-point** type; and (3) the **oblique** or **three-vanishing-point** type. Examples of these are identified in Figure 14.16(a), (b), and (c). Because only drafters in specialty fields are required to produce perspective drawings, the only one that will be discussed here is the angular or two-vanishing-point type, which will provide a working knowledge of technical terms and construction details of how to produce this type of drawing.

14.4.1 The Angular-type Perspective Drawing

In the **angular-type perspective drawing**, the object is positioned at an angle to the projection plane so that two sets of lines converge to two different points called **vanishing points (VP)**, on the **horizon line (HL)**, as represented in Figure 14.16(b). The **ground line (GL)** is the base line or the position where the object rests.

The angular or two-vanishing-point perspective drawing is called angular because of the angle the object makes with the **picture plane (PP)**. All these technical terms will become clear as steps are followed for the construction of this type of technical drawing. The construction process is outlined below and illustrated in Figure 14.17.

Given: The top and side views of object A and the station point (SP) where the observer is positioned (see Figure 14.17).

Step 1: Draw the ground line (GL), horizon line (HL), and picture plane (PP). Locate the horizon line between the top of the side view and the station point.

Step 2: From the station point (SP), project lines which are parallel with the forward edges of object A until they intersect the picture plane (PP). Project lines from these two points vertically downward from the picture plane until they cross the horizon line (HL). These intersections establish the two vanishing points (VP). Drop vertical line XY from the object's point of contact with the picture plane to the ground line (GL). Line XY is parallel with the side view and lies on the picture plane. It appears as a true length in perspective when projected from the side view of object A.

FIGURE 14.15
Oblique ellipses.

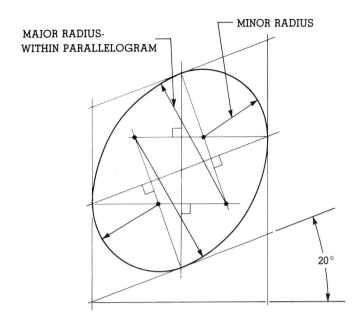

MAJOR RADIUS-
WITHIN PARALLELOGRAM

MINOR RADIUS

20°

LESS THAN 45°

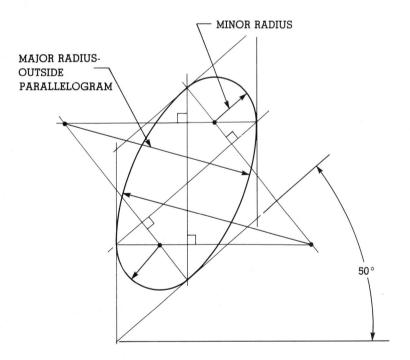

MINOR RADIUS

MAJOR RADIUS-
OUTSIDE
PARALLELOGRAM

50°

GREATER THAN 45°

Step 3: Draw lines from the end points of line XY to the left and right vanishing points (VP) to establish two planes of perspective, as shown. Project lines from the station point (SP) to the three outside corners of object A. From the points where these lines intersect the picture plane, drop vertical lines BC and DE.

Step 4: From the station point, project lines to F and to G. Where these two lines cross the picture plane, draw two vertical lines downward through

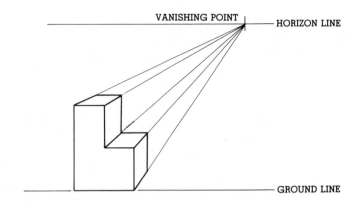

(a) PARALLEL OR ONE-VANISHING-POINT TYPE

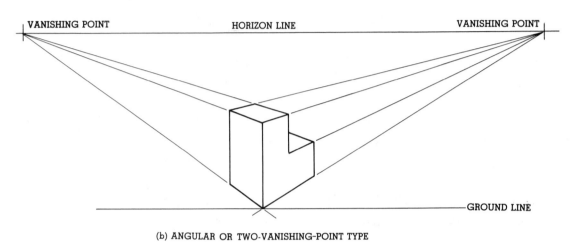

(b) ANGULAR OR TWO-VANISHING-POINT TYPE

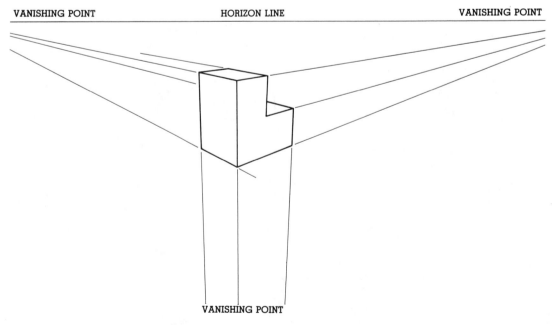

(c) OBLIQUE OR THREE-VANISHING-POINT TYPE

FIGURE 14.16

Types of perspective drawings: (a) Parallel or one-vanishing-point type; (b) Angular or two-vanishing-point type; (c) Oblique or three-vanishing-point type.

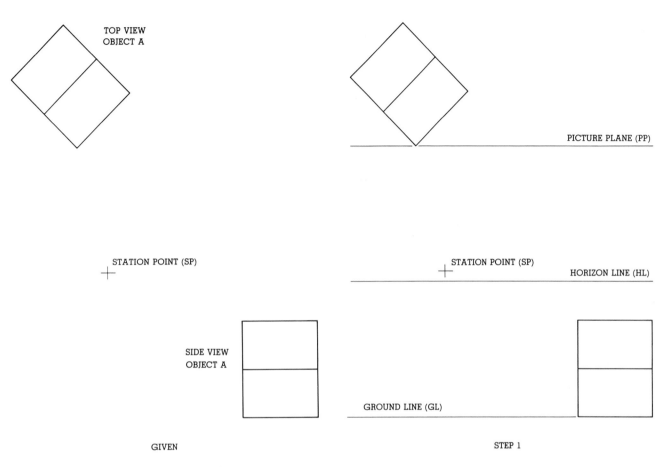

FIGURE 14.17
Procedure for constructing an angular or two-vanishing-point perspective drawing.
(Continued on pages 214–216.)

the perspective view under construction. Project point Z from the side view of object A to the perspective view in process. Complete the construction of the cut-out portion of the object by continuing to the appropriate vanishing points. Complete the remaining visible lines of the perspective drawing. Erase all construction lines and "heavy up" all visible lines. Figure 14.17 illustrates the completed angular or two-vanishing-point perspective drawing.

14.5 SUGGESTIONS FOR PRODUCING THE PICTORIAL DRAWING

Most drawings produced for drafting technology are basically pictorial line drawings, and thus a knowledge of the various types and their applications is important to the drafter. Any of the three basic types of pictorial drawings described in this section may be used for technical illustration purposes, but some may be more suitable than others depending on the

application. The shape of the object to be drawn usually determines which type of pictorial is produced. The isometric type is very simple to draw but is the least natural in appearance. The oblique drawing will be easier to draw if the object is circular or has circular features in the principal view. The perspective drawing is the most realistic in appearance and is free from distortion, but its construction is more time-consuming for the drafter.

The following suggestions are offered for producing pictorial drawings.

1. Select the type of pictorial drawing to be produced: isometric, oblique, or perspective.
2. Determine the medium and tools that will be used. Will vellum, mylar, grid paper, or a sketch pad be required? Decide if templates will be required for drawing regular circles and ellipses in isometric or oblique drawings. Make certain the correct lead and line weight will be used to produce the various lines needed for the drawing.

FIGURE 14.17 (continued)

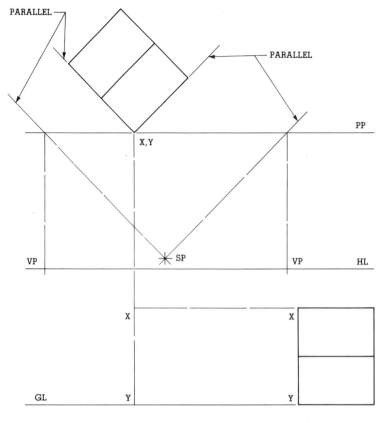

STEP 2

3. Carefully select the orientation in which the object will be drawn so that important features can be identified.
4. Draw the axes. In the isometric drawing, one vertical axis and two axes at 30 degrees from the horizontal will be necessary. For the oblique drawing, front-view lines are drawn to true shape and size. In the perspective drawing, take care that the vanishing points are not so close together that a steep angle of convergence will occur. Try to place them as far apart as the drawing medium will permit.
5. Draw light-weight lines that will box in the entire object, making certain that vertical lines are parallel with the vertical axis and that horizontal lines are parallel in the isometric and oblique drawings but converge in the perspective drawing.
6. A circle or hole may need to be shown as an ellipse that has both a major and a minor diameter where the construction requires a minor and major radius. An ellipse template should be utilized, whenever possible, to facilitate the drawing of an ellipse.

7. Label appropriate points and lines on the drawing with numbers (1, 2, 3, 4) or letters (A, B, C, D) for ease of transferring them or for construction purposes.
8. "Heavy up" all visible lines. Hidden lines are not usually shown.
9. Erase all construction lines and other unwanted lines.
10. Add notes, dimensions, and any other text specifications required to complete the drawing.

14.6 SUMMARY

The pictorial drawing is used for various types of functions by personnel in the industrial and business sectors. The drawings in this category which are most commonly used include the isometric, oblique, and perspective types.

The isometric drawing was defined, and details such as isometric axes, isometric lines, nonisometric lines, isometric curves, isometric circles, and isometric arcs were described.

FIGURE 14.17 (continued)

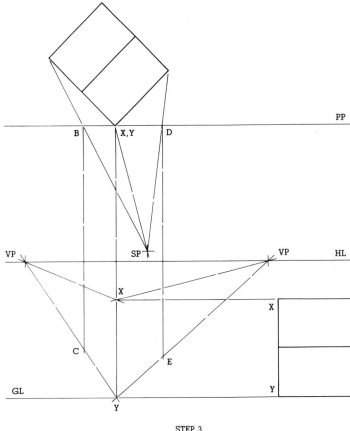

STEP 3

Three kinds of oblique drawings were identified, but only one—the cabinet type—was covered in depth. The oblique drawing was shown to be unique in that the receding axis can face to the left, right, above, or below the front or principal view, and its angle can vary. In addition, the oblique angled surface, oblique circles and arcs, and methods for their construction were illustrated.

The perspective drawing differs from the isometric and oblique drawings in that it represents more realistically what the human eye actually sees. It is used widely by architects in drawings of build-ings, drawings of internal and external views, and landscape and presentation drawings. Three basic types of perspective drawings were identified, but only the angular or two-vanishing-point drawing was covered in detail. Construction aspects and technical terms used for the perspective drawing were described. Knowledge of the various types of pictorial drawings and their application is important to the drafter, and suggestions for producing them were presented. Practice exercises for this section are presented on pages 14A through 14G in the workbook.

FIGURE 14.17 (continued)

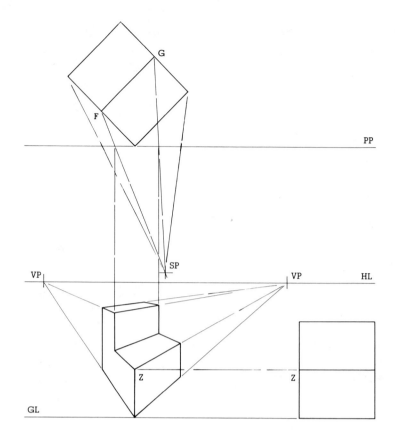

STEP 4

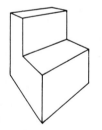

COMPLETED PERSPECTIVE OF OBJECT A

14.7 REVIEW EXERCISES

1. List the most commonly used types of pictorial drawings.

2. Describe an isometric drawing.

3. How many axes are required to produce an isometric drawing? Name them.

4. What is the definition of a nonisometric line?

5. On an isometric drawing, in what form is a circle drawn?

6. In how many different planes can an ellipse be shown?

7. Name three types of oblique drawings.

8. Name the three axes in an oblique drawing.

9. To what scale is the receding axis drawn in a cabinet-type oblique drawing?

10. In what view would it be preferable to show holes or circles on an oblique drawing? Why?

11. How does the perspective drawing differ from the isometric and oblique drawings?

12. Define the following terms used in the construction of perspective drawings:

a. Station point

b. Ground line

c. Vanishing point

d. Picture plane line

13. Name three kinds of perspective drawings.

14. What are the characteristics of an angular or two-vanishing-point perspective drawing?

15. Why is knowledge of the various types and applications of pictorial drawings important to the drafter?

16. Which type of pictorial drawing is the simplest to produce?

17. Which type of pictorial drawing is the most realistic in relation to what the human eye actually sees?

18. Identify six suggestions for producing a pictorial drawing.

19. Which type of pictorial drawing is used extensively in the architectural profession?

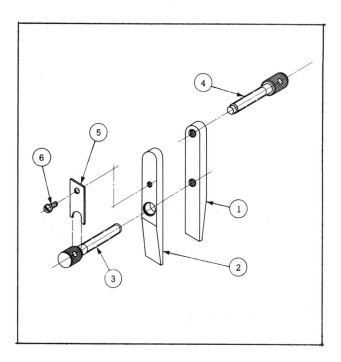

Section **15**

The Assembly and Detail Drawings

LEARNER OUTCOMES

The student will be able to:

- List two groups of working drawings

- Identify ten areas of importance on a drawing format

- Identify seven items of information normally found on an assembly drawing

- Define four types of assembly drawings

- Name two types of detail drawings

- List eight methods of fabricating sheet-metal parts

- Define technical terms used for producing sheet-metal drawings

- Demonstrate learning through the successful completion of review and practice exercises

15.1 PURPOSE

Several kinds of drawings are required to satisfy the needs of business and industry. Among them is a group referred to as **working drawings**. A working drawing is a form of technical drawing that presents information, ideas, and instructions in both pictorial and text forms. Working drawings are usually classified into two groups of releasable drawings: the **detail drawing** and the **assembly drawing**. Both of these drawings are utilized as aids in the manufacture, production, and assembly of parts, machines, and structures.

15.2 PARTS OF A DRAWING

Every drawing should contain general information for the user such as the title of the part, part number, scale, design and drafting approval data, date, drawing field, and so on. The method of presenting and locating this information on a drawing medium varies from one industry to another, but for the purpose of this book the American National Standards Institute (ANSI) specification ANSI Y14.1 will be used.

15.2.1 The Drawing Field

Every drawing has a **drawing field**, which is the main body of the medium and the area to be used for drawing all the necessary views, including dimensions and text, to complete the drawing. An example of the drawing field is illustrated in Figure 15.1. Note that,

depending on the complexity of the object to be drawn, it may be necessary to draw the part off-center on the medium because of other data that may need to be included.

15.2.2 The Title Block

The **title block** is that area of a drawing that contains pertinent information that is not provided on the drawing field itself. It appears in the lower right-hand section of the drawing format and includes individual areas for specific information, as shown in Figure 15.2. Figure 15.2(a) identifies the size of title block used for "A" and "B" size formats, and Figure 15.2(b) shows the recommended title-block dimensions for all other format sizes. The title-block format is usually preprinted on drawing formats of sizes A (8½ by 11 inches) through E (34 by 44 inches).

The various regions, or "blocks," that comprise the title block are as follows:

- **The drawn block** In the drawn block, the drafter's name appears in printed form. In addition, the date of the drawing is shown, usually in the following way: year, month, day (e.g., 87-6-30).
- **The approval block** Some or all of the individual blocks that comprise the approval block are used depending on the complexity of the project, the structure of the organization, and whether it is under contract with the U.S. government or with another company. The "check" block is for the name of the engineering checker who is assigned to the project; the "design" block is for the name of

FIGURE 15.1
The drawing field.

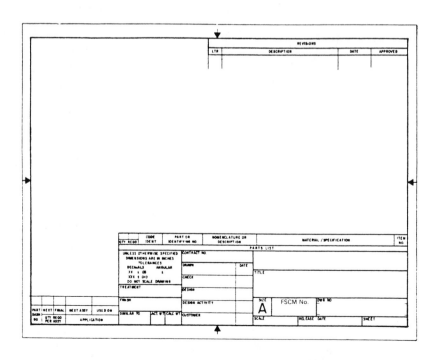

FIGURE 15.2
Recommended dimensions for
title blocks.

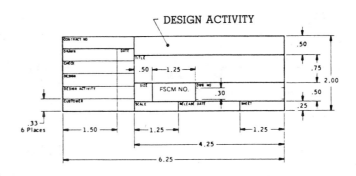

(a) TITLE BLOCK FOR "A" AND "B" SIZES

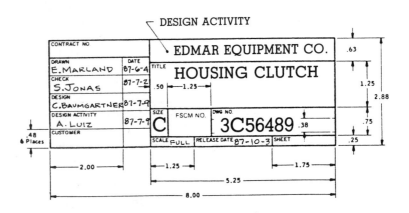

(b) TITLE BLOCK FOR ALL SIZES EXCEPT "A" AND "B"

the responsible engineer; and the "design activity" block is for the name of the program manager or chief engineer. Each block is signed and dated by authorized project or program personnel.

- **The design activity name and address block** This block is for the company name, address, and/or trademark. It is identified, in our case, as EDMAR EQUIPMENT CO.
- **The drawing title block** This area is for the basic name or title of the part or assembly shown in the drawing.
- **The size block** This block contains the letter (A, B, C, D, E, or F) designating the drawing size.
- **The drawing number block** This is where the drawing number for the part is inserted.
- **The scale block** This block is for identifying to what scale the part is drawn (i.e., full, half, etc.).
- **The release date block** When the drawing is complete with all the necessary approvals, the date that the drawing is released to manufacturing is inserted here.
- **The sheet block** Some drawings, because of their complexity, require several pages or sheets.

When this occurs, the sheet number (e.g., 1 of 4) is inserted in this block. If a drawing consists of only one sheet, the sheet block is left blank.
- **The FSCM number block** This block is only for those companies who are under contract to the U.S. government. Each company has an identifying number. FSCM stands for Federal Supply Code for Manufacturers.

In addition to the information inserted in the title block, more data is very often required. Figure 15.3 identifies this supplementary data, which is located to the immediate left of, and borders on, the title block. Only the more important items are detailed below.

- **The standard tolerance block** This area contains standard tolerances that apply to dimensions on the drawing. These tolerances, at times, are uniformly stated for an entire project.
- **The finish block** The finish block is used to specify the type of coating that is to be applied to the part or item. It may be a specification for a protective or decorative coating, or both.

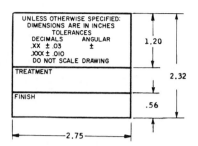

FIGURE 15.3
Supplementary data.

■ **The treatment block** Information on heat treatment, temper, or hardness, usually in the form of a specification number, is inserted here, if required.

The above information may seem complex, but nonetheless it is used by many companies and is required for work completed under contractual agreement with the U.S. government and its agencies.

In the format used for the practice exercises in the workbook, the title block and approval areas are quite simple. This has been done as a result of space limitations and in an effort to maximize the field of the drawing. The drawing format that will be used is shown in Figure 15.4.

15.2.3 The Revision Area

Many drawings will require one or more changes during their lifetime—changes in dimensions, shapes, or forms due to changes in design or to correct errors in design or drafting. To accommodate engineering changes on drawings, the **revision area** (or **revision block**) has been established. It is normally preprinted as part of the drawing format. Dimensions of revision blocks recommended for "A" and "B" size formats, and for formats of other sizes, are shown respectively in Figure 15.5(a) and 15.5(b).

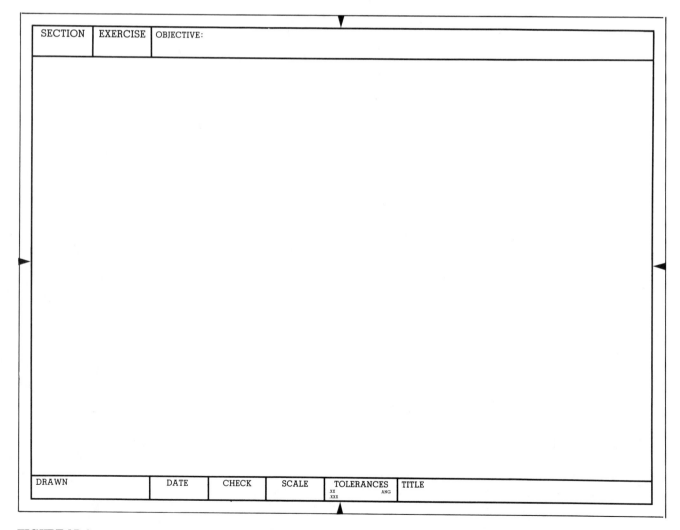

FIGURE 15.4
Drawing format for practice exercises.

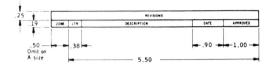

(a) REVISION BLOCK FOR "A" AND "B" SIZES

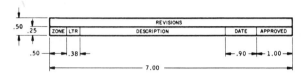

(b) REVISION BLOCK FOR ALL SIZES EXCEPT "A" AND "B"

FIGURE 15.5
Recommended dimensions for revision blocks.

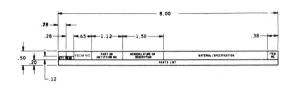

(a) PARTS LIST FOR "A" AND "B" SIZES

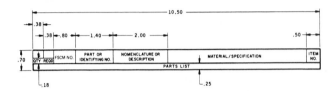

(b) PARTS LIST FOR ALL SIZES EXCEPT "A" AND "B"

FIGURE 15.6
Recommended dimensions for parts lists.

According to American National Standards Institute (ANSI) recommendations, the revision area should be located in the upper right-hand corner of the drawing and should provide space for the description of the change, the revision symbol zone location, the date of the change, and approval of the change.

15.2.4   The Parts List

The **parts list** is actually a list of materials which identifies those items which are required to complete an assembly or all the necessary parts of some functional grouping. The parts list includes the required quantity of each part, the part or identifying number, the title or description of the part, the material of the part, and the item number of each part. Sample parts lists are presented in Figure 15.6. The parts list is located in the lower right-hand corner of the drawing format above the title block. When the parts list is an integral part of the drawing, the lettering is done by the drafter. If the parts list is not part of the drawing, it can either be lettered by the drafter or take the form of a computer printout.

15.2.5   The Numbering System

Every company has some form of **numbering system** for recording and identifying drawings. There is no single best system to use. Most industries use a system in which drawing numbers are assigned sequentially as each drawing is completed. Others assign an entire block of numbers to a project, especially if it is

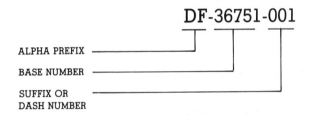

WHERE MORE THAN ONE ITEM IS DESCRIBED ON A DRAWING, DIFFERENTIATE AMONG THEM BY ASSIGNING EACH A UNIQUE AND CONSECUTIVE DASH NUMBER AS FOLLOWS:

A812090-015
A812090-016
A812090-017
A812090-018

FIGURE 15.7
Example of a numbering system.

a large project. At times a prefix or suffix letter is added to the number of the drawing. Some companies use as little as four digits in their numbering system while others use as many as 15. An example of a numbering system is presented in Figure 15.7.

15.3   THE DESIGN LAYOUT DRAWING

Development of the **design layout drawing** is usually the responsibility of the designer or design engineer.

FIGURE 15.8
Examples of design layout
drawings.

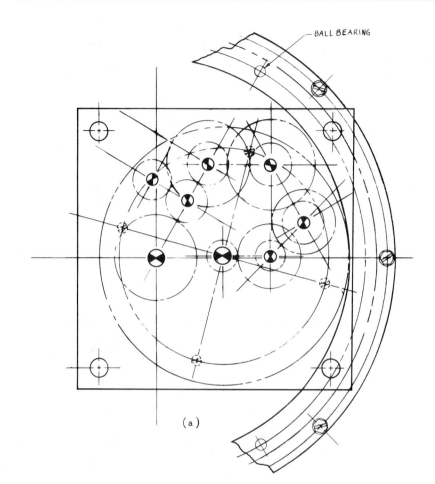

(a)

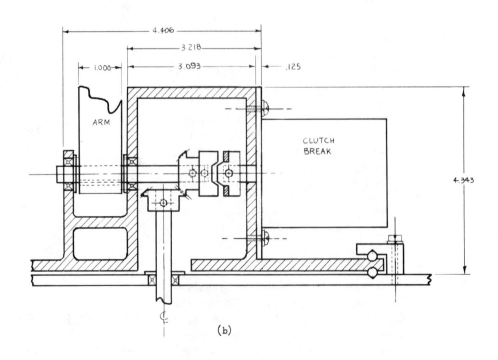

(b)

It is a drawing that graphically illustrates a design concept and is the origin or basis for all detail and assembly drawings. The design layout drawing is not used for purchasing of parts, for inspection, or for manufacturing purposes, but rather presents sufficient information to the detail drafting group so that a complete set of engineering drawings may be completed with a minimum amount of consultation with the designer.

Layouts are drawn accurately and to scale so that detail drawings may be prepared by scaling the layout drawing. Layouts that are of full size or larger are preferred. For purposes of accuracy, a layout is seldom drawn less than full scale. For example, if a layout is drawn at one-quarter scale, any error due to shrinkage of the drawing medium will be magnified four times. For this reason, plastic film or some other suitable stable material should be utilized as the drawing medium for the design layout drawing.

In the process of producing the layout, care should be taken to ensure that all necessary design information, such as dimensions, tolerances, clearances, materials, finishes, and processes critical to the design, is included so that the detail drafter will have a minimum number of questions relative to the design. Examples of design layout drawings are shown in Figure 15.8.

15.4   THE DETAIL DRAWING

After the design layout drawing has been completed and approved, the drafting group is called upon to produce the **detail drawing**, which means that each and every individual part must be "picked off" the layout and drawn. The detail drawing presents all the information necessary for fabricating an item, including the **shape description**, the **size description**, and the **specifications** of the part.

The shape description consists of information which describes or shows the shape of the part. The size description shows the locations and sizes of features on the part. Specifications include text information such as material, finish, and heat treatment data.

Two types of detail drawings are utilized by industry. One is called the **mono-detail drawing** and the other the **multi-detail drawing. Mono** implies that only one detail drawing is produced for each part. **Multi** means that two or more detailed parts may appear on a single sheet of drawing medium. Figure 15.9 illustrates the differences between mono-detail and multi-detail drawings.

15.5   THE ASSEMBLY DRAWING

Functional groupings of components for machines and mechanisms for consumer, industrial, and military products are composed of numerous parts. A drawing depicting a product in its completed form is called an **assembly drawing**. An assembly drawing represents, through the technology of drafting, the assembled relationship of (1) two or more parts; (2) a combination of detail parts which are joined to form a subassembly or a complete unit; or (3) a group of assemblies required to form an assembly of higher order. Assembly drawings differ from one another depending on their complexity and the information required, but normally contain the following:

1. A sufficient number of views to clearly show the relationship among parts
2. Sectional views necessary to show internal features, function, and assembly
3. Enlarged views to identify necessary details
4. Correct arrangement of parts
5. Principal reference dimensions or dimensions critical to assembly
6. A parts list including reference to parts through the use of balloon number identifiers
7. Necessary manufacturing processes required for assembly

15.5.1   Types of Assembly Drawings

Of the several different types of assembly drawings that drafting personnel are asked to produce, the four most common are the **separable, inseparable, detail**, and **expanded** assembly drawings.

- **The separable assembly drawing**   This type of drawing shows the assembly relationship between two or more items where at least one of the items can be disassembled for servicing or replacement without causing damage or destruction to any of the other items. This drawing is also referred to as a **general assembly drawing** and is, by far, the most common. Figure 15.10 shows an example of a separable assembly drawing.
- **The inseparable assembly drawing**   The inseparable assembly drawing represents the assembled relationship between two or more parts, separately fabricated, but permanently joined together by brazing, riveting, cementing, welding, or soldering, and not subject to disassembly. The assembly is considered as a single item. Figure 15.11 depicts a drawing of an inseparable assembly.

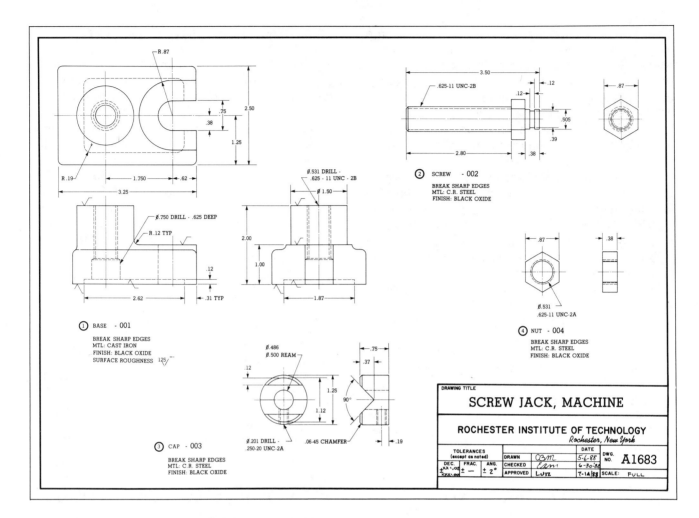

MULTI-DETAIL DRAWING

FIGURE 15.9
Multi-detail and mono-detail drawings.

■ **The detail assembly drawing** Figure 15.12 illustrates the detail assembly drawing. Note that this type of drawing has several dimensions because, while it identifies the relationship between assembled items, it also details one or more parts of the assembly. Individual drawings are not required for items so illustrated, but fabrication details should be an important consideration.

■ **The expanded assembly drawing** This drawing is also referred to as an **exploded view drawing**. It graphically describes an assembly/disassembly relationship, in either isometric or perspective form, among the parts of an assembly that appears to have "exploded" along center lines. This type of drawing frequently appears in vendors' catalogs as well as in instruction or maintenance manuals for consumer, industrial, and military

products. An expanded assembly drawing appears in Figure 15.13.

15.5.2 Part Identification

On an assembly drawing it is absolutely critical that some method of relating the items on the parts list with those shown graphically on the drawing be utilized. The primary method for such keying is the use of **balloon numbers**, which consist of identification numbers placed inside circles called balloons. Each balloon is attached to a related part with a leader line. The leader line is a thin line that usually terminates in an arrowhead (or solid dot) that points to, and touches, the related part on the field of the drawing, as shown in Figure 15.14. Note that balloon numbers are not placed at random, but rather have

FIGURE 15.9 (continued)

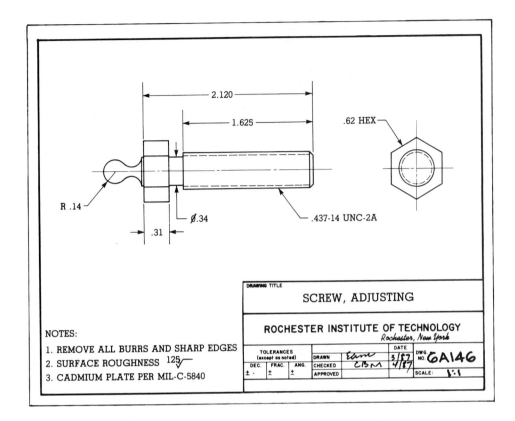

NOTES:
1. REMOVE ALL BURRS AND SHARP EDGES
2. SURFACE ROUGHNESS 125
3. CADMIUM PLATE PER MIL-C-5840

DRAWING TITLE

SCREW, ADJUSTING

ROCHESTER INSTITUTE OF TECHNOLOGY
Rochester, New York

TOLERANCES (except as noted)			DRAWN	Eam	DATE 3/87	DWG. NO. 6A146
DEC. ± .	FRAC. ±	ANG. ±	CHECKED	CBM	4/87	
			APPROVED			SCALE: 1:1

MONO-DETAIL DRAWING

FIGURE 15.10
The separable assembly
drawing.

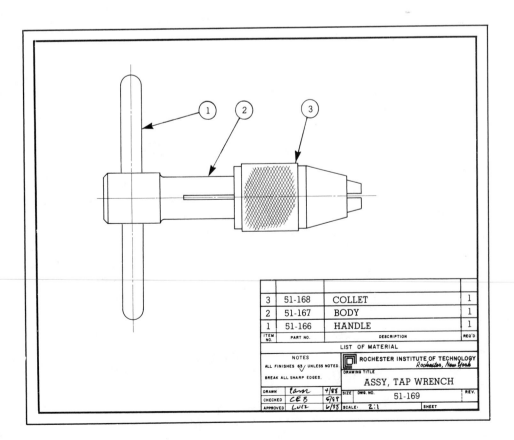

3	51-168	COLLET	1
2	51-167	BODY	1
1	51-166	HANDLE	1
ITEM NO.	PART NO.	DESCRIPTION	REQ'D

LIST OF MATERIAL

NOTES
ALL FINISHES 63 UNLESS NOTED.

BREAK ALL SHARP EDGES.

ROCHESTER INSTITUTE OF TECHNOLOGY
Rochester, New York
DRAWING TITLE
ASSY, TAP WRENCH

DRAWN	Eam	4/88	SIZE	DWG. NO. 51-169		REV.
CHECKED	CEB	5/88				
APPROVED	Luiz	6/88	SCALE: 2:1		SHEET	

FIGURE 15.11
The inseparable assembly drawing.

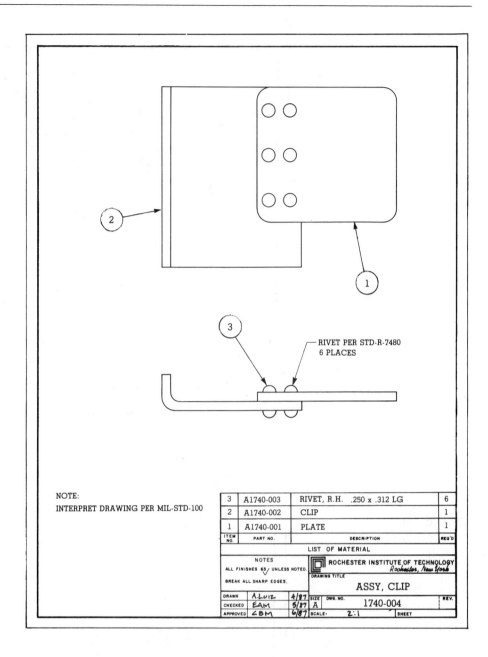

NOTE:
INTERPRET DRAWING PER MIL-STD-100

RIVET PER STD-R-7480
6 PLACES

ITEM NO.	PART NO.	DESCRIPTION	REQ'D
3	A1740-003	RIVET, R.H. .250 x .312 LG	6
2	A1740-002	CLIP	1
1	A1740-001	PLATE	1

LIST OF MATERIAL

NOTES
ALL FINISHES 63/ UNLESS NOTED.
BREAK ALL SHARP EDGES.

ROCHESTER INSTITUTE OF TECHNOLOGY
Rochester, New York

DRAWING TITLE
ASSY, CLIP

DRAWN	A.Luiz	4/87	SIZE	DWG. NO.		REV.
CHECKED	EAM	5/87	A	1740-004		
APPROVED	CBM	6/87	SCALE:	2:1	SHEET	

some order—either a horizontal or vertical alignment, or both. Balloons vary from .31 to .75 inch in diameter depending on the size and scale of the assembly drawing.

15.6 THE SHEET-METAL DRAWING

Equipment enclosures and mountings for mechanical and electromechanical components can take many forms. Very often these components are mounted on or in equipment such as a **rack**, a **chassis**, or a **cabinet** or **console**. This equipment is produced from thin-gage metal called **sheet metal**. The fabrication of sheet-metal parts includes practices which are quite different from those used for machined parts, and it is necessary for the drafter to be knowledgeable about these differences so that technical drafting preparation is of good quality.

Sheet-metal parts are usually produced by **shearing, bending, drawing, spinning, trimming, punching, perforating, blanking**, or any combination thereof. There are several technical terms peculiar to this topic with which the drafter should be familiar. They include:

■ **Rack** A thin-gage metal structure that includes long vertical members (angles). It is normally made of steel or aluminum and houses electronic, electromechanical, or mechanical equipment,

FIGURE 15.12
The detail assembly drawing.

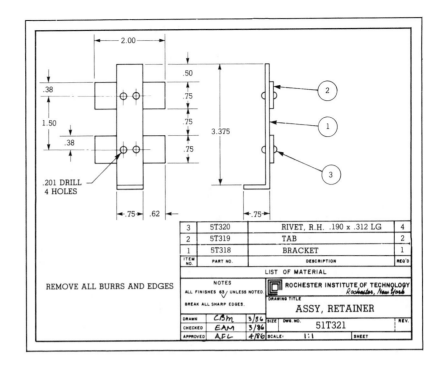

FIGURE 15.13
The expanded assembly
drawing.

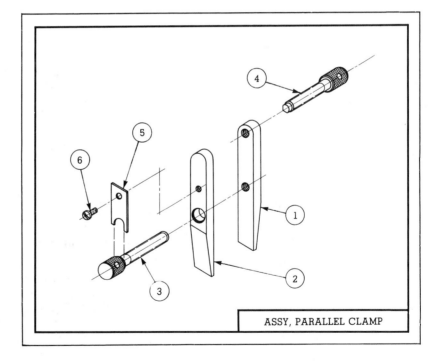

either permanently mounted or mounted in drawers with slides. A standard rack is illustrated in Figure 15.15.

■ **Chassis** A thin-gage metal structure or base that is designed to support electronic, electromechanical, or mechanical devices. It may be of any size or shape and can be mounted on a standard rack panel or used separately. It may or may not have a cover or be enclosed. Examples of chassis are shown in Figure 15.16.

■ **Cabinet** Normally an aluminum or steel structure of either standard or special shape, depending on the application. The horizontal mountings are designed to accept a standard rack-mounted panel.

FIGURE 15.14
Part identification with balloons.

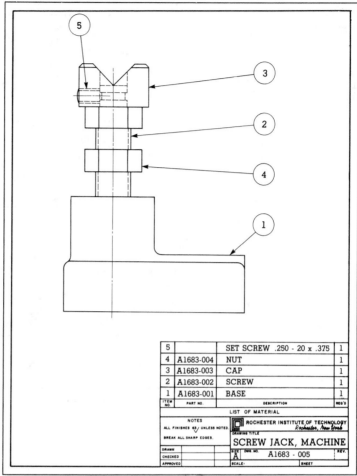

5		SET SCREW .250 - 20 x .375	1
4	A1683-004	NUT	1
3	A1683-003	CAP	1
2	A1683-002	SCREW	1
1	A1683-001	BASE	1
ITEM NO.	PART NO.	DESCRIPTION	REQ'D

LIST OF MATERIAL

NOTES		ROCHESTER INSTITUTE OF TECHNOLOGY	
ALL FINISHES 63 / UNLESS NOTED.		*Rochester, New York*	
BREAK ALL SHARP EDGES.		DRAWING TITLE SCREW JACK, MACHINE	
DRAWN		SIZE A	DWG. NO. A1683 - 005	REV.
CHECKED			
APPROVED		SCALE:	SHEET

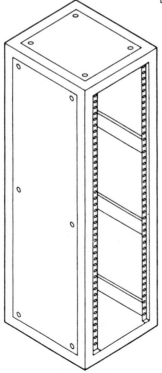

FIGURE 15.15
Standard equipment rack.

A cabinet with a desk-top surface may also be referred to as a **console**. Figure 15.17 depicts a cabinet.

- **Bend allowance** The length of material around a bend from bend line to bend line, as represented in Figure 15.18.
- **Bend angle** The angle to which a piece of sheet metal is bent. It is measured from the flat through the bend to the finished angle after bending, as pictured in Figure 15.19.
- **Bend line** The line of tangency where a bend changes to a flat surface, as depicted in Figure 15.18. Note that each bend has two bend lines.
- **Bend radius** The minimum inside radius to which the material can be bent without fracturing or cracking at the outside radius. The bend radius is identified in Figure 15.19.
- **Blank** A piece of flat sheet-metal stock of the size required to make a formed sheet-metal part. Before sheet metal is formed it is referred to as being **in the flat**.
- **Leg** The flat or straight section of a part after bending or forming. Two legs are shown in Fig-

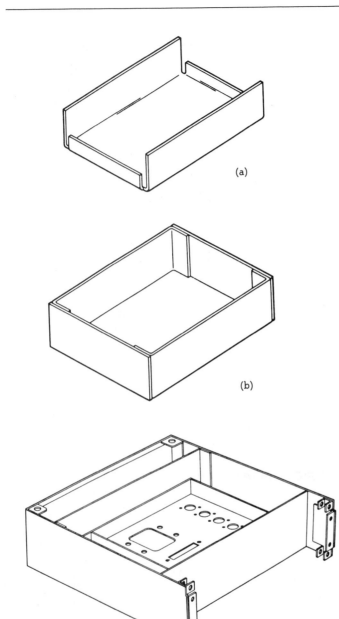

FIGURE 15.16
Common chassis: (a) U-type; (b) Box type; (c) Rack mounted type.

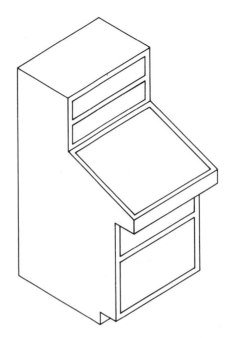

FIGURE 15.17
Cabinet or console.

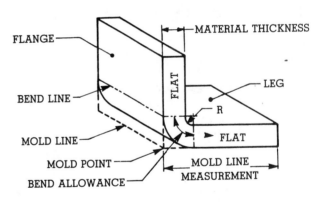

FIGURE 15.18
Sheet-metal terms.

ures 15.18 and 15.19. Legs are also referred to as **straight lengths**.

■ **Mold line** The line of intersection of two outside flat surfaces of a formed sheet-metal part, as illustrated in Figure 15.18.

15.6.1 Developed Length

One of the most common operations performed on sheet metal is forming or bending. This operation usually takes place after all holes, slots, and cutouts

are produced in the blank, or when the part is in the flat. Bending can be done on a hand or power-operated machine called a **brake**. The quantity of items to be produced determines whether a hand brake or a power-operated brake is used. The radius of a bend varies depending on the type and thickness of the material. If too small a radius is specified on a drawing, the material may develop cracks or actually fracture at the bend.

When designing an electronics, electromechanical, or mechanical package of sheet metal, care should be taken that the appropriate bend radius is specified and that the resulting **developed length** is correct. Appendix J may be used for refer-

FIGURE 15.19
Bend angle.

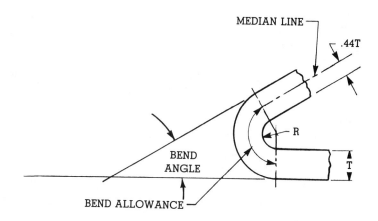

ence in determining the appropriate bend radius for a given application.

Developed lengths and widths need to be accurate so that material dimensions can be determined and the department producing the part will have an idea as to the initial size of the stock to be used. Appendix J may be referred to for solutions to problems which require the calculation of developed lengths with 90-degree bends or greater.

15.6.2 Designing Sheet-metal Parts

The cost of sheet-metal parts varies with the quantity to be produced and the design of the part. Small quantities dictate the use of low-cost tooling such as **punches, dies, jigs, fixtures**, and low-production-type machines. As a result, the unit cost is usually high. Attention to the considerations in the list that follows will contribute to the design of more economical products.

1. Whenever possible, a design should make use of standard and available tooling.
2. Try to utilize minimum numbers of different hole sizes and patterns, bend radii, bend reliefs, and inside or outside bend radii in the design.
3. Considering all factors, select a design approach which calls for uniform and minimum stock thickness.
4. Avoid tolerances which are closer than necessary. Economical design means using tolerances which are as generous as practical.
5. Use progressive dimensioning from datum lines whenever possible, since this system locates part features with respect to datum lines and not with respect to each other. Figure 15.20 illustrates progressive dimensioning of a flat part as well as dimensioning of features related to each other.
6. On flanged or formed parts, locate the datum points on the lower and left-side edges of the part, as shown in Figure 15.21. Try to locate the

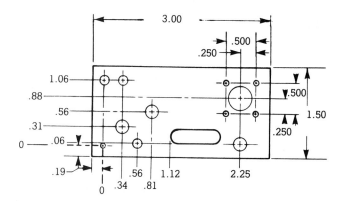

FIGURE 15.20
Dimensioning of a flat sheet-metal part.

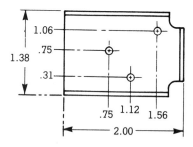

FIGURE 15.21
Dimensioning of a formed sheet-metal part.

datum points on an unformed edge.

7. Locate round holes by dimensions between hole centers. Noncircular cutouts are located by dimensioning to the cutout feature which is most important to the function of the item, as depicted in Figure 15.22.
8. Whenever sheet-metal bends intersect, such as at a corner, it is necessary to remove material from the intersection area to prevent wrinkling or tearing of the material during the forming

FIGURE 15.22
Locating holes and noncircular cutouts.

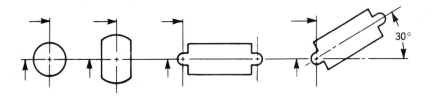

FIGURE 15.23
Bend-relief cutouts.

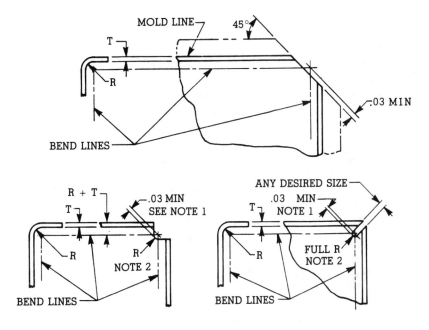

NOTES:
1. PREFERRED LOCATION FOR CENTER OF CUTOUT RADIUS IS THE INTERSECTION OF THE BEND RADIUS.
2. MINIMUM RADIUS IS THE RADIUS OF THE SMALLEST PERMISSIBLE HOLE WHICH MAY BE PUNCHED IN THE SPECIFIED MATERIAL.

operation. Material must be removed to a minimum of .03 inch beyond the intersection of bend lines. There are several configurations for a **bend relief**, three of which are illustrated in Figure 15.23. Preferred practice is to radius the inside of a relief cutout, but a sharp corner is acceptable provided that material is removed as shown in the figure.

9. Raised or depressed areas in sheet-metal parts, usually provided for the purpose of rigidity, are called **beads**. The top view of the bead illustrated in Figure 15.24 shows the outline of the bead at the mold line in phantom lines. In addition, sections or breakouts are required for dimensioning the radii and depth of the bead. When a straight section is required between the

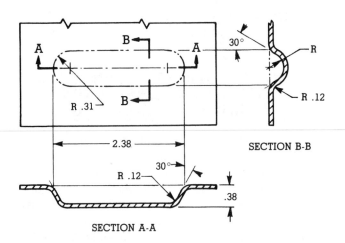

FIGURE 15.24
Dimensioning of beads.

FIGURE 15.25
Minimum flange height, hole spacing, and edge distance.

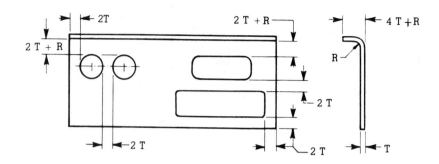

two bend radii, the angle must also be dimensioned.

10. **Minimum flange height, minimum hole spacing**, and **minimum edge distance** are all illustrated in Figure 15.25. The minimum flange height that should be considered is four times the material thickness plus the bend radius. Minimum hole spacing—that is, the minimum distance between the edge of a circular or noncircular punched hole and any other hole or edge—should be twice the material thickness. The edges of holes or openings adjacent to a bend should be a minimum distance of twice the material thickness plus the bend radius away from the inside surface of the flange. The minimum diameter for punched holes in steel or aluminum is equal to the material thickness. Smaller holes must be drilled holes.

11. **Drawn parts** are parts which require the stretching or compressing of material. They are dimensioned to either the outside or inside mold lines depending on design requirements. All dimensions are given to the same side of the material except bend radii, which are always given to the inside of the bend. Tolerances on drawn parts should not be less than ±.03 inch, and larger tolerances are preferred. Figure 15.26 illustrates design and dimensioning of drawn parts.

15.7 SUGGESTIONS FOR PRODUCING DETAIL, ASSEMBLY, AND SHEET-METAL DRAWINGS

When a product is developed and drawings are required it is beneficial to know a few basic steps for producing the technical drafting involved. Most entry-level drafters will be called upon to produce detail drawings, relatively simple assembly drawings, and sheet-metal drawings that are not too complex. The following suggestions are offered for producing each of these types of drawings.

15.7.1 The Detail Drawing

1. Using a sketch pad, draw all expected views freehand. Draw at full scale, if possible, so that the proper size of drawing medium can be selected.

2. There is no need to place dimensions on the sketch at this point, but allow sufficient space to accommodate all expected dimensions.

3. When satisfied that the proper size and type of medium has been selected, draw the basic outline of the detail on the medium selected using light-weight lines. Be certain that the area on the

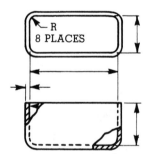

DIMENSIONS TO INSIDE MOLD LINES

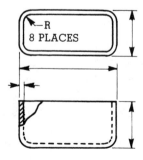

DIMENSIONS TO OUTSIDE MOLD LINES

FIGURE 15.26
Design and dimensioning of drawn parts.

field of the drawing used will not interfere with the revision block area.

4. Leave space for text and general notes.

5. After the detail drawing is complete and the proper line weights are in place, make a white print or use whatever type of copier is available to make a copy of the drawing, before dimensions are added.

6. Add dimensions to the copy of the detail drawing. When satisfied as to the number and location of dimensions, transfer the information to the original drawing.

7. Add all text, hole-size, and thread information if required. Add specifications, if needed. Fill in the title block, the tolerance block, and the approval block.

8. Review the drawing for accuracy, completeness, and neatness.

15.7.2 The Assembly Drawing

1. Gather all the detail drawings, standard hardware catalogs, and all other documentation needed to produce the assembly drawing.

2. Determine the complexity of the assembly drawing to be produced. Ask several questions, such as:
   a. What is the total number of parts that comprise the assembly?
   b. How many views will be required?
   c. Should one or more of the views be in section? Should one or more views be enlarged?
   d. Is a **breadboard** or an **engineering model** of the assembly available as an aid to clear visualization of the parts?
   e. What is the most appropriate scale to use?

3. What type and size of medium will be used? Be certain to select one that will accommodate not only the views required but also space for the parts list and for possible revisions.

4. After the medium has been selected, draw the assembly, clearly showing any details that will facilitate the actual assembly of parts.

5. Show only those hidden lines which are needed for clarity. The accepted practice is to draw exterior views to clarify outside features and sectional views to define interior features. Do not section line standard hardware parts.

6. Add dimensions only to those parts whose assembly is critical or must conform to specific fits.

7. Add a parts list including part numbers, numbers of parts required, descriptions of parts, hardware, and item numbers.

8. Add balloon numbers (with leader lines and proper line terminations).

9. Review for completeness and accuracy.

15.7.3 The Sheet-metal Drawing

The general guidelines for producing the detail drawing also apply to the sheet-metal drawing with the exceptions identified below:

1. Determine the type of dimensioning system to be used (i.e., chain, progressive, datum, etc.) for the part.

2. Be sure to keep economics in mind when producing this type of drawing. Try to specify holes, slots, and cutouts that can be produced with existing tools.

3. Will the drawing be a detail drawing or an inseparable assembly drawing? If the latter, what process of joining will be used (i.e., brazing, soldering, welding, bonding, etc.)?

4. What mechanical operations best suit the part (i.e., bending, forming, blanking, perforating, beading, etc.).

5. Determine the number of views needed, including sections, breakouts, etc.

6. Adhere to recommended design practices (see Subsection 15.6.2).

7. Add material and finish information to the drawing.

8. Review for completeness and neatness.

15.8 SUMMARY

The purpose for the existence of detail and assembly drawings was introduced. Details of the various parts of a drawing were discussed, including the field of the drawing, the title block, the drawn block, the approval block, the design activity and address block, the drawing title block, the size block, the drawing number and scale blocks, and the release date and sheet blocks. Supplementary data for standard tolerance, finish, and treatment information was also defined. The revision area, the parts list, and numbering systems for drawings were discussed.

Several types of drawings were illustrated and covered in depth, including the design layout drawing, the detail drawing, the assembly drawing, and the sheet-metal drawing. It was emphasized that sheet-metal parts are produced by various fabrication methods including bending, shearing, drawing, spinning, trimming, punching, perforating, and/or blanking. Several technical terms specifically used in the design of sheet-metal parts were identified and defined, and information on calculation of developed lengths was introduced.

Design information for producing economical sheet-metal parts was presented, and suggestions were offered for developing and drawing detail, assembly, and sheet-metal drawings. Practice exercises for this section are presented on pages 15A through 15J in the workbook.

15.9 REVIEW EXERCISES

1. What is a working drawing?

2. List ten important components of a drawing format.

3. What is the drawing field used for?

4. Give three examples of supplementary data that may be required on a drawing.

5. What part of a drawing format is used to record changes in the drawing?

6. What is the purpose of the parts list on an assembly drawing?

7. What is used to record and identify drawings within a company?

8. Define the term "design layout drawing."

9. What kind of information does a detail drawing provide?

10. Explain the differences between mono-detail and multi-detail drawings.

11. What does an assembly drawing represent?

12. List four types of assembly drawings. Define each.

13. List three types of mounting equipment which are made from thin-gage metal.

14. Why are accurate developed lengths of sheet-metal parts important?

15. Identify six items to consider when designing an inexpensive sheet-metal part.

16. List five suggestions for producing a detail drawing.

17. List five suggestions for producing an assembly drawing.

18. List four suggestions for producing a sheet-metal drawing.

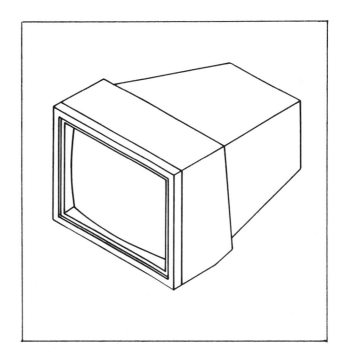

Section **16**

An Introduction to Computer-Aided Drawing

LEARNER OUTCOMES

The student will be able to:

- State the purpose of a CAD system

- List the major hardware elements of a computer-aided drafting (CAD) system and describe their functions

- Give several examples of the input device, the output device, and the processing device

- Describe the differences between hardware and software

- Identify the three types of CAD systems and their differences and similarities

- Define several technical terms associated with CAD systems

- Produce several types of technical diagrams and drawings utilizing CAD system hardware and software

- Demonstrate learning through successful completion of review and practice exercises

16.1 INTRODUCTION

This section on computer-aided drawing has not been placed here as an add-on or an afterthought. The specific rationale for placing it at this point in the textbook is the belief that the drafter should possess an appropriate background in basic manual technical drafting development before proceeding into the world of automated drafting. With a good grounding in basic manual drafting skills, the drafter should be able to move quickly and intelligently into the field of computer-aided drawing.

The purpose for all the learning experiences in this course up to this point has been for the learner to obtain basic manual skills in drafting technology by (1) gaining knowledge of standard drawing practices and drafting laboratory procedures and (2) acquiring the skills to produce quality mechanical technical drawings.

Computer-aided drafting (CAD) systems have been used by government agencies and by industry for several years, providing new and imaginative ways of producing drawings. The primary purpose of a CAD system is to increase productivity by enabling the drafter/designer to make neater, cleaner, more accurate drawings in less time and to have parts manufactured directly from automated graphic part information. A CAD system is an automated electronic tool that replaces traditional equipment and tools. The key to any CAD system is the operator—in our case, the drafter. Computer-aided drafting systems are being used in all engineering fields, in some to a greater degree than in others. One field where CAD is used extensively is the production of mechanical diagrams and drawings.

Acronyms are used throughout the computer graphics industry by various companies that design and manufacture computer graphics systems and equipment. An **acronym** is a word formed by using first letters of a title or phrase. Some of the more common acronyms which describe these systems include:

- **CAD** Computer-Aided Design/Drafting; Computer-Assisted Design/Drafting
- **CADD** Computer-Aided Design and Drafting; Computer-Assisted Design and Drafting
- **CAE** Computer-Aided Engineering
- **CAA** Computer-Aided Artwork
- **CAG** Computer-Aided Graphics
- **ADS** Automatic Drafting System
- **EGS** Engineering Graphics System

16.2 THE BASIC CAD CONCEPT

Let us begin by considering a simplified version of a basic computer-aided drafting (CAD) concept: first,

the computer receives information (**input**) from the **keyboard** or **digitizer**; second, it processes that information and displays the result (**output**) on a **monitor** or **plotter**. For the graphics system, the computer uses **data** (the input) provided by the operator (drafter) to produce (by processing the data) a drawing (the output). Figure 16.1 illustrates a typical computer-aided drafting concept.

CAD systems are generally **interactive**, which implies that an interaction or communication takes place among the various components of the system as well as between the system and the operator. The types of drawings that can be produced accurately and rapidly for industries that produce products of a mechanical nature include, but are not limited to, those identified in Section 1, page 2.

Although one must go through a time-consuming learning process before acquiring proficiency in the use of CAD **hardware** and **software**, some claim that productivity by engineering personnel can be increased by a factor of 4 to 1 over conventional or traditional practices for design and drafting, depending on the type of CAD system and the number of peripheral options. The learning curve for those in training to become CAD operators varies according to many factors, but a period of three to six months of training on a CAD system usually allows the operator to become proficient.

Relatively high costs have prevented educational institutions from procuring large **mainframe** or **turnkey** CAD systems, but several firms have recently entered the market with moderate-to-low-cost systems that use **minicomputers** or **microcomputers**, which are desktop computers with workstations and **user-friendly**, intelligent terminals.

Many secondary and postsecondary institutions have been able to work out arrangements with large mainframe manufacturers whereby these schools have received CAD hardware at reduced cost in addition to receiving accompanying software

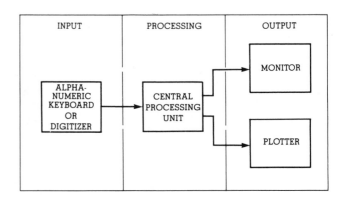

FIGURE 16.1

The basic CAD concept.

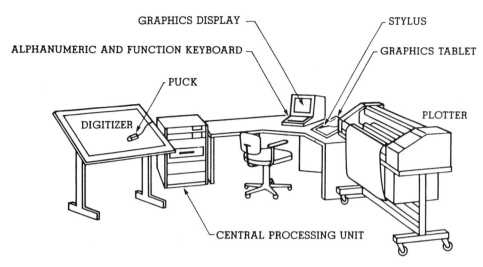

GRAPHICS DISPLAY

ALPHANUMERIC AND FUNCTION KEYBOARD

STYLUS

GRAPHICS TABLET

PUCK

PLOTTER

DIGITIZER

CENTRAL PROCESSING UNIT

FIGURE 16.2
Two-dimensional interactive CAD system.

at no cost. We will describe several CAD systems in the sections that follow and will define their structures, capabilities, similarities, and differences.

16.3 SYSTEM COMPONENTS

Computer-aided drafting systems consist of two basic elements, hardware and software. **Hardware** may be defined as the physical components of the system, including electrical, electronic, electromechanical, magnetic, and mechanical devices. **Software** consists of sets of procedures, programs, and related documentation that direct operation of the system to produce graphics and related text material.

Hardware for a basic CAD system always consists of a computer that includes a **central processing unit (CPU)**, a **memory device**, and a **storage facility** for programs and files. A typical CAD workstation may also consist of a **digitizer** and **puck**, a **graphics tablet** and **stylus**, a **function keyboard**, a **graphics display**, and a **plotter** for producing **graphical output**. The system depicted in Figure 16.2 is a complete interactive system for **two-dimensional** drafting. Examples of software include information contained in memory, on disks, in files, in the CPU, and in storage.

16.3.1 The Input Device

The **input device** provides data and instructions to the computer. Every input device takes data from the operator and prepares it for processing by the central processing unit (CPU) of the CAD system. Computer-aided drafting systems use various means and types of input devices for processing information. Several in use are identified below:

- **Alphanumeric keyboard** The alphanumeric keyboard is normally similar to a typewriter keyboard, but may also include special function keys. It allows the user to input letter ("alpha") and number ("numeric") instructions to the CPU. It may be used as a programming device or as a word-processing tool. An illustration of the alphanumeric keyboard is presented in Figure 16.3.
- **Cursor** This is not really an input device but assists in providing input. It determines the placement of the next character on the screen in the form of a bright marker. It may be variously shaped as a dot, crosshair, check, or rectangle. An example of a cursor is shown in Figure 16.4.
- **Digitizer** The digitizer is an electronic tracing table or board which is used to enter existing drawings into the CAD system by means of a pointing device and a menu. Figure 16.5 depicts a digitizer.

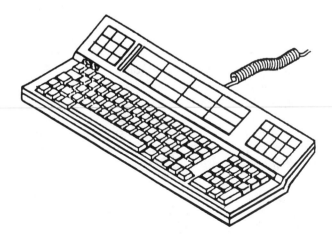

FIGURE 16.3
Alphanumeric keyboard.

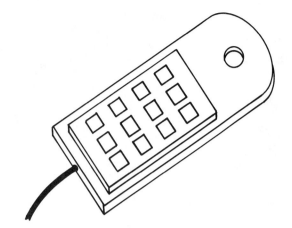

FIGURE 16.4
Cursor.

FIGURE 16.6
Graphics tablet.

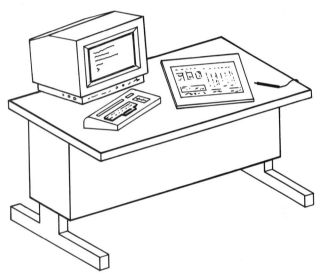

FIGURE 16.5
Digitizer.

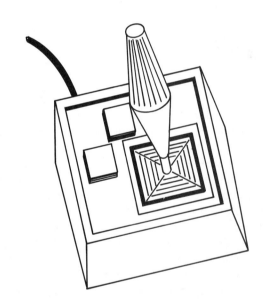

FIGURE 16.7
Joystick.

■ **Graphics tablet** This is a device which has a flat surface on which work is performed. A **stylus**, **puck**, or "**mouse**" is used for graphical data entry, and information is transmitted to the cathode-ray tube (CRT) by means of an electronically controlled grid beneath the tablet's surface. Figure 16.6 shows a typical graphics tablet.

■ **Joystick** This device consists of a lever in a weighted base. Movement of the lever (stick) controls the screen's cursor. The user moves the lever in the same direction as he or she wishes the cursor to move on the screen. Figure 16.7 shows a joystick.

■ **Light pen** With a light pen the user can point directly at an area of the graphics monitor screen to enter points or select commands from a screen menu. Crosshairs on the screen follow the pen un-

til the user selects a point or menu item by releasing or deactivating the pen. Operation of a light pen is shown in Figure 16.8.

■ **Menu tablet** The menu tablet has a flat surface on which functions are selected by the user with a stylus or puck. This device is also known as a **menu pad**. A typical menu tablet is displayed in Figure 16.9.

■ **Modem** The modem is both an input and an output device because it is capable of sending (inputting) and receiving (outputting) information. The

FIGURE 16.8
Light pen. (Courtesy of
CADAM, Inc.)

FIGURE 16.9
Menu tablet.

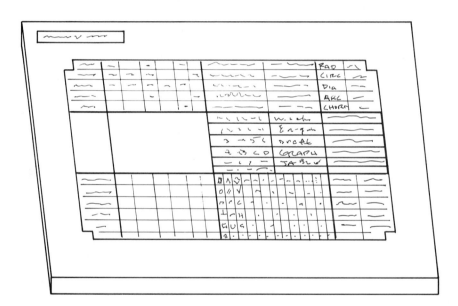

vehicle for the transmission of data is a regular telephone line. Modems are available in various **baud** (speed) rates. A modem is shown in Figure 16.10.

- **Mouse** A mouse is made operational by moving it around a flat surface while crosshairs track its movement on the screen. To activate the point or menu item at which the crosshairs are positioned, the user presses a button on the mouse. Figure 16.11 shows a typical mouse input device.

- **Prompt** The prompt is usually an arrow ($>$), asterisk ($*$), or dot ($\cdot$) at the beginning of a line on the screen, signifying that the system is ready for the user to input a command. Examples of prompts are shown in Figure 16.12.

- **Puck** The puck is a hand-held device which is movable on a graphics tablet. It has essentially the

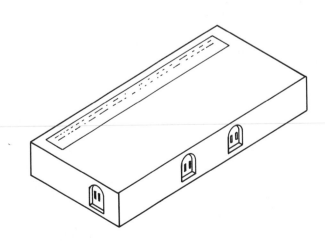

FIGURE 16.10
Modem.

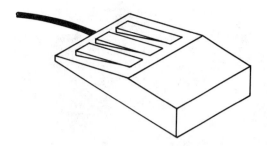

FIGURE 16.11
Mouse.

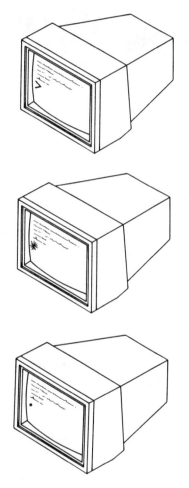

FIGURE 16.12
Prompts.

FIGURE 16.13
Puck.

the face of the monitor. The user selects a point or menu item by sliding the pen tip along the panel until the crosshairs are positioned at the desired point, then lifting the pen from the screen. An example of this device is presented in Figure 16.15.

■ **Voice-data entry** This system activates verbal input commands by the user. It is currently in use and allows the user to keep eyes and hands free to perform other tasks, which results in fewer errors.

16.3.2 The Output Device

The reason for inputting information and data into the computer is to receive **output**. The output takes the form of either "visual copy" or "hard copy." **Visual copy** is output observed by the eye on the graphics screen. **Hard copy** is data produced by a printer or a plotter, on paper or film. The **output device** is the vehicle which produces the output. The major output devices are as follows:

■ **CRT** The CRT (cathode-ray tube) projects electrons onto a fluorescent screen which glows when the electron beams are excited, and as a result produces a graphics display on the video screen. A CRT is depicted in Figure 16.16.

■ **Graphics display** This unit displays the image, drawing, or text material. The most popular graphics display is the CRT. The terms "monitor," "CRT," "screen," and "display" are used interchangeably with the term "graphics display." An example of a graphics display is shown in Figure 16.17.

■ **Plotter** This output device is a piece of equipment that draws either mechanically (with a pen)

same effect as a mouse. A puck is shown in Figure 16.13.

■ **Screen menu** On some CAD systems a menu is displayed on the graphics screen. The menu allows input commands to be entered simply by pointing to the command on the display screen. A screen menu is illustrated in Figure 16.14. This is commonly referred to as a "pull-down" menu.

■ **Touch pen** This device operates much like a light pen but is operated by touching a special panel on

FIGURE 16.14
Screen menu.

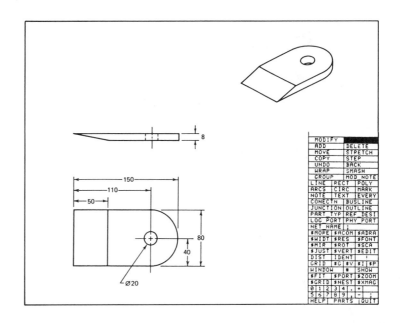

FIGURE 16.15
Touch pen.

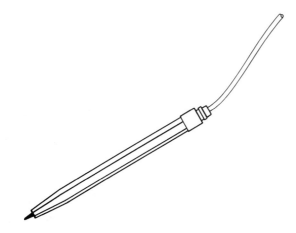

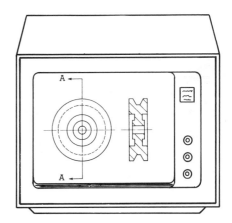

FIGURE 16.17
Graphics display.

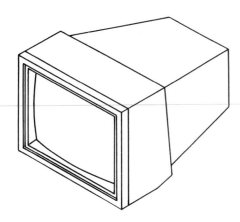

FIGURE 16.16
CRT.

or electrostatically on paper, vellum, or polyester film and produces hard copy. Examples of such devices include the flatbed, drum, and electrostatic plotters. Figure 16.18 shows these three types of plotters.

■ **Printer** The printer is an output device that is capable of duplicating the image of a graphics display. One example is an alphanumeric unit similar to a typewriter. Because the printer does not produce high-quality hard copy, its use is limited to production of text and preliminary or check print copies rather than of permanent graphic output. Devices of this type include the **dot-matrix**, **line**, and **laser** printers. Figure 16.19 shows examples of printers used on CAD systems.

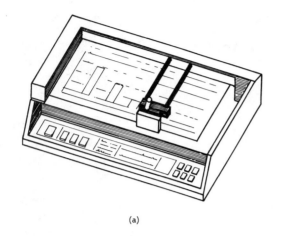

(a)

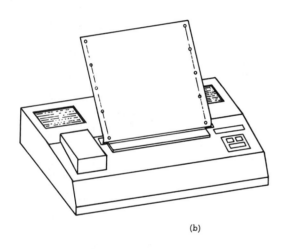

(b)

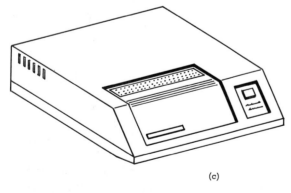

(c)

FIGURE 16.18
Plotters: (a) Flatbed plotter; (b) Drum plotter; (c) Electrostatic plotter.

16.3.3 The Processing Device

As indicated in Figure 16.1, the three major areas of the CAD concept are input, output, and processing.

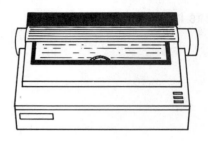

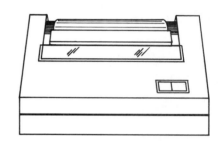

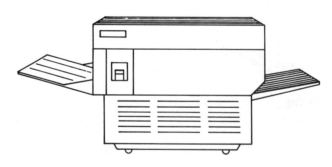

FIGURE 16.19
Printers.

Processing devices are those units which process input information and data. After input is inserted into the system, the data is processed—manipulated, if you will—before it appears as output. Processing devices not previously covered under input and output (I/O) devices are identified below:

- **Central processing unit (CPU)** The CPU is considered to be the brain of the computing system. **Integrated-circuit (IC) chips** form an important part of the microprocessor portion of the computer, which accomplishes the logical processing

FIGURE 16.20
CPU.

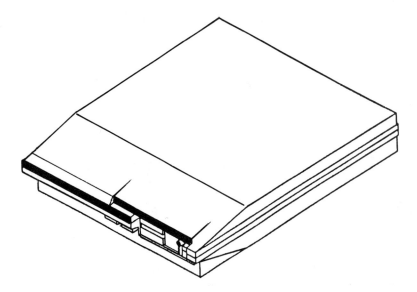

of data. The CPU contains the arithmetic, logic, and control circuits and often incorporates memory storage. Figure 16.20 shows a CPU.

- **Memory** Storage of data is a function of memory. The size of a memory is determined by the amount of information it can store. The smallest unit of memory is the **bit**. The bit, designated by a **binary number** (a 0 or a 1), may or may not contain an electrical charge. A **byte** equals eight bits. A CAD system may contain several thousand (**K**) to several million (**mega**) bytes of memory.

- **Disk drive** A disk drive is a device that retains, reads, and writes onto disks. Most systems utilize two disk drive units. A disk drive is the equipment that operates the software packages. Figure 16.21 depicts a typical disk drive unit for a microcomputer-based system.

- **Disk** Disks are available in two forms and in different sizes. The **floppy disk** is a portable, flat, black or brown circular plate made of thin vinyl with a magnetic coating that is inserted into and removed from the disk drive unit. It is a software package of stored data, housed in standard square sizes of 3½, 5¼, or 8 inches. The **hard disk**, sometimes referred to as the **Winchester disk**, is a nonremovable disk with a large capacity for storing information in the form of files. Its size depends on the amount of memory that is desired or that will be necessary to operate the system successfully. Information is easily transferred between floppy disks and hard disks.

16.4 THE MAINFRAME-BASED SYSTEM

The physical configuration of a computer-aided graphics system can take several forms. The most versatile, powerful, and expensive form is the **mainframe-based system**. These systems are capable of performing numerous tasks and functions across a wide range of complex engineering areas. The term **mainframe** means that the CAD system consists of a CPU with a large capacity for design and analytical tasks that may accept the workloads of several **workstations** (or **terminals**); each workstation has input and output capability but the CPU is at a remote location. In most cases, several terminals operate off one main or **host computer**: this is called "networking" or "time sharing."

A mainframe system which is in wide use by research organizations, governmental agencies, industrial corporations, and educational institutions is the **Intergraph Interactive Graphics System**. Specific forms of this system, which are produced by the Intergraph Corporation, are the InterAct and the InterPro 32. These are complete computer systems, which implies that they can be used as stand-alone systems or as terminals in a computer network. Either the InterAct or the InterPro 32 can also be used as:

- A graphics terminal for running Intergraph graphics applications on a VAX or Micro II computer

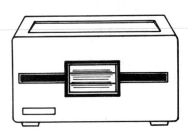

FIGURE 16.21
Disk drive.

- A general-purpose workstation for running UNIX applications
- A general-purpose VT-220 terminal for running VAX/VMS applications on an Ethernet network
- A personal computer for running IBM-PC-compatible applications

Figure 16.22 is a graphic representation of the hardware for these two systems.

16.4.1 The System Hardware

The InterAct and InterPro 32 terminals include the following components.

16.4.1.1 The Alphanumeric Keyboard

The keyboards on both units are identical. This low-profile keyboard is similar to a standard typewriter with the addition of a numeric key pad, modifier keys, screen function keys, and programmable function keys. Figure 16.23 illustrates the low-profile keyboard.

The keys that are most concerned with graphics operations are the RETURN, DELETE, HOLD SCREEN, and CONTROL keys. The numeric key pad at the right of the keyboard is an alternative method of entering numbers. The functioning of these keys is

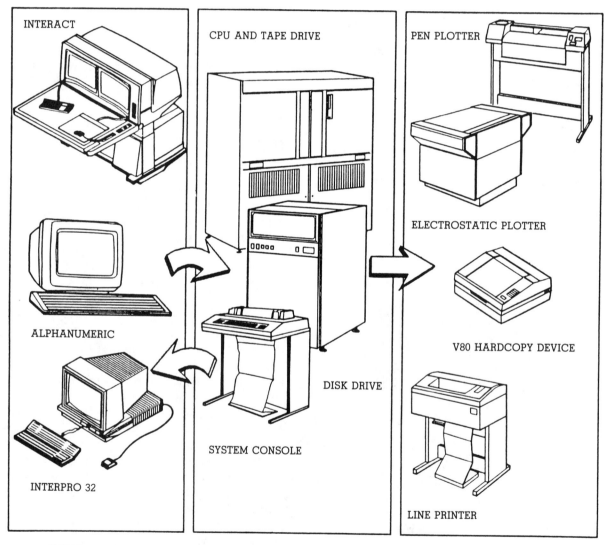

INTERACT

CPU AND TAPE DRIVE

PEN PLOTTER

ALPHANUMERIC

ELECTROSTATIC PLOTTER

INTERPRO 32

DISK DRIVE

SYSTEM CONSOLE

V80 HARDCOPY DEVICE

LINE PRINTER

INPUT/OUTPUT   PROCESSING AND STORAGE   OUTPUT

FIGURE 16.22
Intergraph Interactive Graphics System hardware.

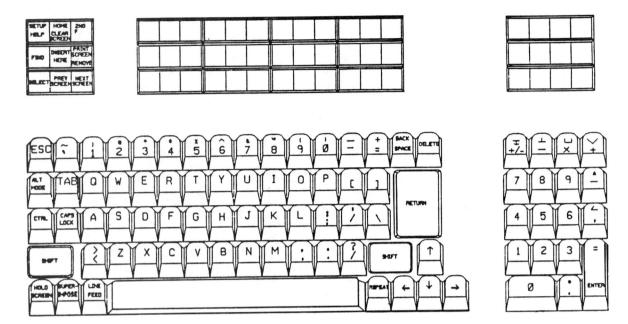

FIGURE 16.23
Low-profile alphanumeric keyboard for InterAct and InterPro 32.

exactly like that of the numeric keys on the main part of the keyboard.

The terminal function keys are located immediately above the main keyboard and include the PRINT SCREEN, SELECT, and NEXT SCREEN keys.

16.4.1.2 The Input/Output (I/O) Devices

Input/output devices include the InterAct and InterPro 32 graphics terminals (shown in Figure 16.24) and the alphanumeric terminal, which is used only for general textual input. These devices allow the operator both to input information to the computer and to receive information via the central processing unit (CPU). The graphics terminals accept both drafting input and data keyed in from the keyboard, whereas the alphanumeric terminal accepts only keyed-in data.

The InterAct graphics terminal permits access to the graphics system. It is used to enter design drawings and information, retrieve the designs for update or correction, and perform certain other tasks. It has two screens, a keyboard, and a handheld cursor, and uses a paper menu.

The InterPro 32 terminal, primarily an engineering workstation, provides full access to Intergraph applications. It can be used as a stand-alone computer or as a terminal in a network. It has one screen, a keyboard, and a mouse, and uses screen pull-down menus.

16.4.1.3 The Processing and Storage Devices

The major processing and/or storage devices include the CPU, the magnetic tape drive, the disk drive, and the system console.

- **Central processing unit** and **tape drive** The CPU controls the overall activity of the system, processing all information. Programs are run by the CPU and, depending on user responses to these programs, output is produced. The magnetic tape drive, usually located in or adjacent to the CPU cabinet, resembles a reel-to-reel tape recorder. It both reads and writes data at very high speeds. The CPU can store information on this tape for future use or can transfer the data to another computer. Its main use is to recover data in the event the system does not retain data because of a malfunction.
- **Disk drive** The disk drive, located adjacent to the CPU, functions like a huge filing cabinet. It stores information on system disks in the form of files that users create and assign names to. The CPU manipulates files on the disk as requested. The disk drive also reads and writes data stored on disks. Each disk has a device name by which it is referred to in the process of performing tasks.
- **System console** The system console, which is connected to the CPU, has an alphanumeric keyboard for input and records output on a printer. This device provides access to the system.

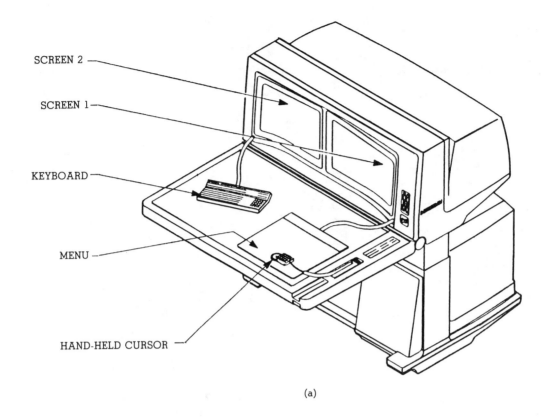

SCREEN 2

SCREEN 1

KEYBOARD

MENU

HAND-HELD CURSOR

(a)

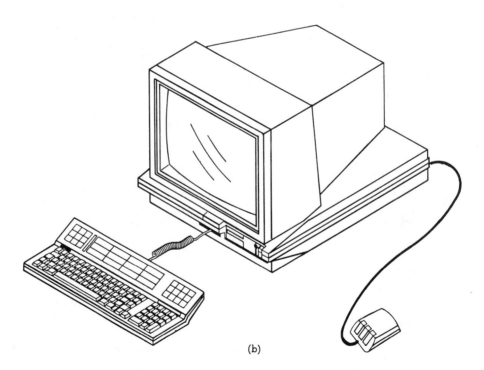

(b)

FIGURE 16.24
Input/Output devices: (a) InterAct terminal; (b) InterPro
32 terminal.

16.4.1.4 Output Devices

Output devices, which provide paper copies of files, include plotters and printers.

- **Line printers** Line printers are used to print alphanumeric output such as a program listing and reports. Line printers with graphics capability can also reproduce copies of CRT images.
- **Plotters** Various types of plotters are used to generate paper copies of graphic information. Three of the most common types are the **electrostatic plotter**, the **pen plotter**, and the **V-80 hard copy device**. The electrostatic plotter uses a liquid toner to produce graphic images in either color or black and white. The pen plotter uses ballpoint or felt-tip pens to produce plots. The V-80 is a small electrostatic plotter that is usually located beside the graphics terminal. It outputs quick copies of graphic files and can also be used to print a hard copy of a current screen image.
- **Display screens** The graphics terminal screens are views into the user design file and can be directed to look at any portion of a drawing. Up to eight sections of a design can be viewed at one time. The InterAct terminal has two physical screens: the right screen (screen 1) and the left screen (screen 2). Although the InterPro 32 has only one physical screen, it emulates the dual-

screened InterAct terminal by having two "virtual" screens, either of which can be displayed on the single physical screen. However, these virtual screens can be viewed only one at a time.

- **Menus** The Intergraph Graphics Design System (IGDS) command menu, which is represented in Figure 16.25, contains commands that support the design process. The hand-held cursor (InterAct) and mouse (InterPro 32) are used to select commands from the menu. In effect, these commands tell the system what is to be accomplished. Many command items on the menu have technical illustrations of the operation of the command and show the numbered data-point entry sequence to be used when executing the command. Other command items are identified only by the command name. The InterAct terminal has a paper menu that is located on the terminal table. The top portion of the menu is a reference for key-ins that must be entered from the terminal keyboard. The rest of the menu consists of the IGDS commands. The InterPro 32 terminal uses screen menus only. There are two types of these menus: The **control strip** menu and the **command** menu, which are shown in Figure 16.26. The control strip menu, which appears at the top of the screen, is used to activate commands that allow for viewing of de-

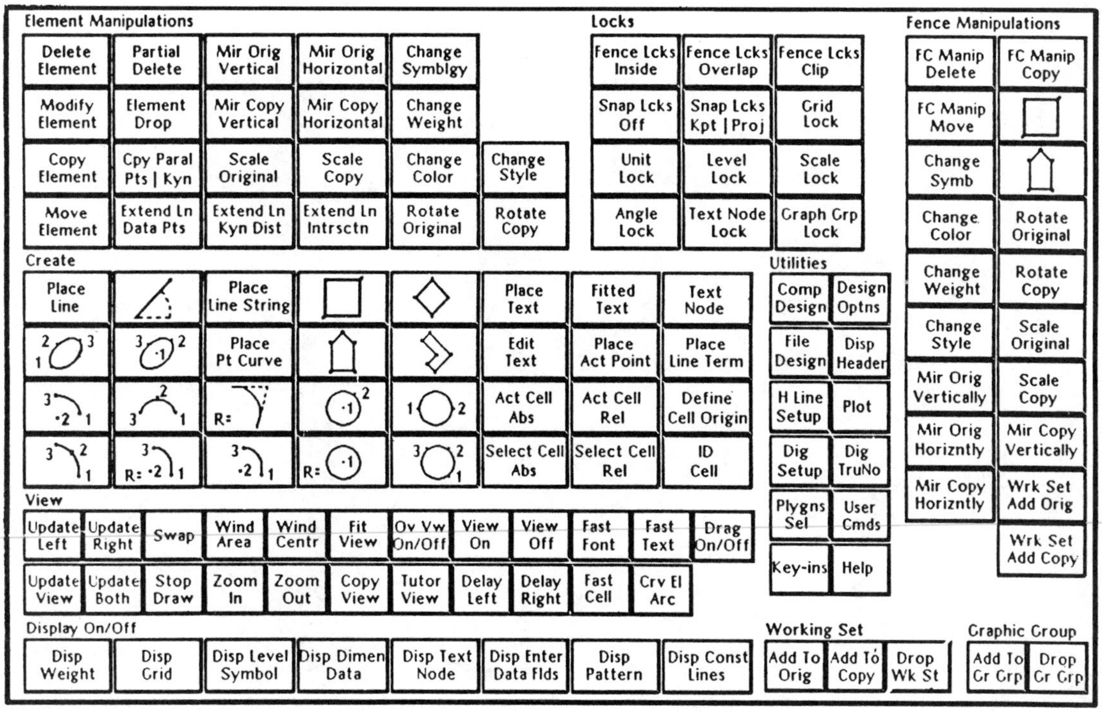

FIGURE 16.25
IGDS command menu.

		ACTIVE MENU					**Actem**
		MENU ACTIVATED					
Create	Meas/Manip	Reference File	Dimensioning	Text	DMRS	TDS	

(a)

View

Update Left	Update Right	Swap	Wind Area	Wind Centr	Fit View	Ov Vw On/Off	View On	View Off	Fast Font	Fast Text	Drag On/Off
Update View	Update Both	Stop Draw	Zoom In	Zoom Out	Copy View	Tutor View	Delay Left	Delay Right	Fast Cell	Crv El Arc	

(b)

FIGURE 16.26
Menus: (a) Control strip menu; (b) Portion of command menu.

sign files, setting of parameters, and manipulation of files. The command menu is used to activate element creation and manipulation commands.

- **Hand-held cursor and mouse** When designing on the InterAct or InterPro 32, the operator will use the hand-held cursor or the mouse, respectively, to perform such functions as selecting a

command from the menu or placing data points in a design. The hand-held cursor used with the Inter-Act has 12 buttons and a crosshair, as demonstrated in Figure 16.27(a). The mouse (similar to the cursor) has only three buttons with no crosshair and is depicted in Figure 16.27(b). When the cursor is on the menu tablet, a corresponding

FIGURE 16.27
Cursor and mouse: (a) Hand-held cursor; (b) InterPro 32 mouse.

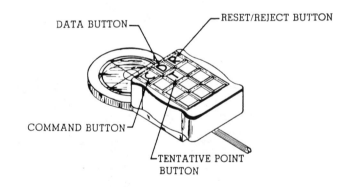

(a)

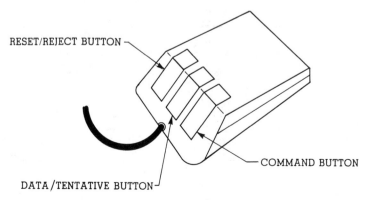

(b)

cross is displayed on the screen (visual cursor). As the hand-held cursor is moved, the visual cursor on the screen moves accordingly. The movements of the mouse are also tracked by a cross on the screen. Examples of the types of technical drawings that can be produced quickly and efficiently on the Intergraph Interactive Graphics System are demonstrated in Figure 16.28.

FIGURE 16.28
Examples of technical drawings produced on the IGDS system.

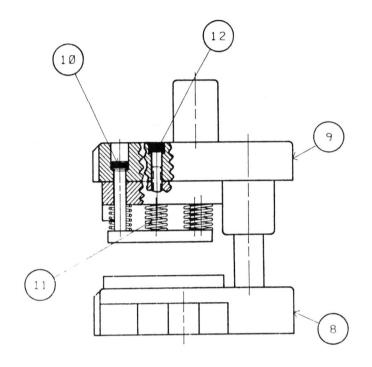

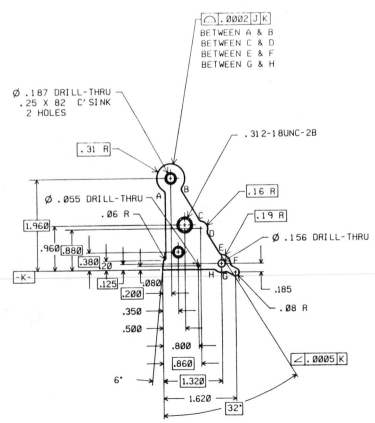

16.5 THE MINICOMPUTER-BASED SYSTEM

In most cases the **minicomputer-based system** would be more appropriately identified as a midsize system because its use falls somewhere between the microcomputer and mainframe systems. The midsize system, until very recently, was probably the CAD system most widely used by small-to-medium-size engineering-related companies and educational institutions. These systems are affordable, simple to operate, and serve a market that does not necessarily require an extremely quick response time. That is, the processor speed is not as fast as that of a mainframe-based system. They are, however, particularly useful for mechanical engineering, design and drafting, and documentation applications where the varied assembled configurations are based on the user's needs. Today's 16-bit and 32-bit computers are capable of supporting multiple workstations whose software has two-dimensional and three-dimensional capability.

16.5.1 The HP ME Series 10

A two-dimensional minicomputer-based CAD system is shown in Figure 16.29. This is the Hewlett-Packard Design Center Mechanical Engineering Series 10 (ME Series 10). It is a versatile design and drafting system for optimizing each step of the design process. It offers opportunities for creative initial design, rapid design modifications, and parametric design, and also features a comprehensive set of commands to simplify drafting and documen-

FIGURE 16.29
Hewlett-Packard Design Center Mechanical Engineering Series 10. (Courtesy of Hewlett-Packard.)

tation tasks.

The power of the ME Series 10 is based on its advanced internal data structure, which results in precise drawings. The system is entered through a friendly and easy-to-learn user interface which uses a combination of tablet and screen menus. A desired command is selected from the tablet or screen menu field using the pen attached to the tablet. The use of each command is self-explanatory with prompts requesting the appropriate user interaction. Figure 16.30 depicts the tablet menu for the ME Series 10, while Figure 16.31 displays the screen menu.

16.5.1.1 Viewports

The hardware of the ME Series 10 allows the operator to display multiple views of a drawing using **viewports**. Lines can be drawn from one viewport to another with changes occurring automatically in all viewports. This makes it possible to work on a detail while still displaying the entire drawing in another viewport. Examples of viewports are illustrated in Figure 16.32.

It is possible to view as many as 16 ports simultaneously. Other viewing modes include Windowing, Zoom in, Zoom out, Pan, and Show (the Show mode exhibits all views of the drawing, including text, crosshatching, etc.).

16.5.1.2 Drafting and Documentation

The ME Series 10 contains all the features needed to produce high-quality drawings rapidly. Drafting standards include ANSI, ISO, DIN, conversions, and company-specific drafting criteria. Measurements

FIGURE 16.30
ME Series 10 tablet menu.

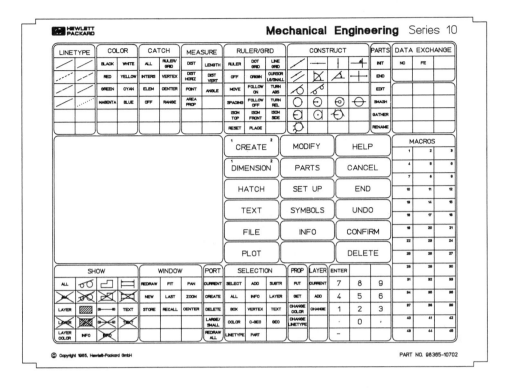

FIGURE 16.31
ME Series 10 screen menu.

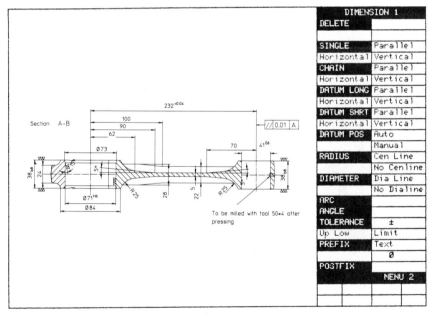

can be expressed in microinches, mils, inches, feet, yards, miles, metric units, or user-defined units.

A ruler capability emulates the T-square used on a drafting board. Several modes are available to allow fast sketch input and easy isometric or orthogonal construction. The equipment also provides a fully comprehensive set of construction geometry commands whereby direct construction of the final drawing elements is also possible, including lines, rectangles, circles, arcs, fillets, chamfers, splines, ellipses, and equidistant contours.

This system offers a large number of dimensioning commands tuned to the requirements of drafting technology. Default values can be set to meet various dimensioning standards, and dimensioning is updated automatically when the geometry is modified. Commands in the dimensioning mode include horizontal, vertical, parallel, chain, datum, coordinate, radius, diameter, arc, angle, prefix, postfix, tolerances, imperial, metric, dual, fractional, and unit conversions. In addition, crosshatching is performed quickly and easily and is updated automati-

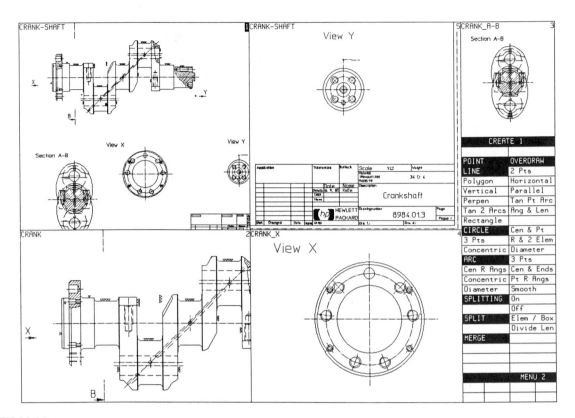

FIGURE 16.32
Viewports of ME Series 10.

cally when the geometry is modified. Outstanding text and symbols facilities are provided by the ME Series 10, and drawing information can be structured using an unlimited number of layers.

Design features are instrumental in relieving users of routine and repetitive tasks. A drawing or part of a drawing may be copied, rotated, mirrored, scaled, or stretched. Grouping and nesting of parts simplify creation of complex drawings and simulate the structure of the assembly. Several instances of the same part can occur on one drawing (shared part), such that when the part is altered, all instances of that part are also updated. Parts and assembly commands include the following: create part, create assembly, subassembly, smash part, assembly, detailing functionality, view part, parent assembly, edit assembly hierarchy, parts list assembly tree structure, and parts list count. An example of an assembly drawing and parts list produced on the ME Series 10 is shown in Figure 16.33.

The ME Series 10 is supported by the HP-9000 Series 300 Engineering Workstations under the Pascal or the HP-UX operating system. The Pascal workstations can be networked using HP's Shared Resource Management (SRM) system, and HP-UX terminals can communicate via HP's Local Area Network (LAN).

16.5.1.3 The Plotter

Production of a hard copy of the drawing occurs at the plotter. Three different sizes of plotters are available for this system, each of which has its own special features. The plotters will plot on single sheets from size "A" (8½ x 11 inches) to roll sizes depending on the model of plotter. Plotting is quick and precise, and lines and curves are sharply defined. Resolution on a plotter can be defined as the smallest move that can be specified programmatically. The resolution on each model is 0.025 mm (0.000984 inch). High repeatability allows the plotter to join new lines smoothly to previously plotted ones. The drafting plotter holds eight pens of three types: fiber tip, roller ball, and drafting. Each pen type is mounted in a carousel, as pictured in Figure 16.34. When not in use, pens remain tightly capped in the carousel to prevent premature ink drying. Four buttons control normal plotter operation, as shown in Figure 16.35. These buttons are used for medium (paper, mylar) loading, and tell the plotter to begin accepting program instructions from the computer. They allow interruption of plotting so the operator can view the drawing as it is being produced. The front panel also provides local control of many plotter functions. The full front panel is illustrated in Figure 16.36.

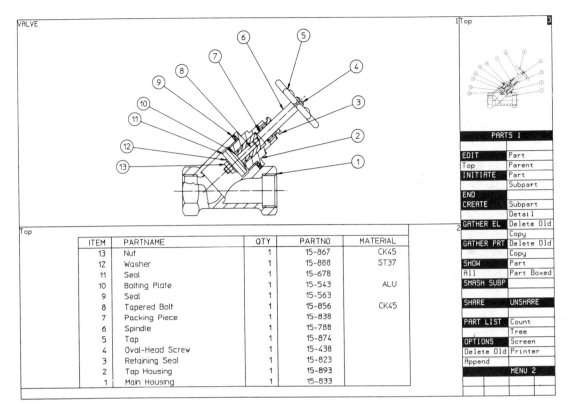

FIGURE 16.33
Assembly drawing and parts list produced on ME Series 10.

FIGURE 16.34
Pen carousel. (Courtesy of Hewlett-Packard.)

FIGURE 16.35
Plotter control panel. (Courtesy of Hewlett-Packard.)

FIGURE 16.36
Full control panel. (Courtesy of Hewlett-Packard.)

16.6 THE MICROCOMPUTER-BASED SYSTEM

Several **microcomputer-based systems** are available in the marketplace. They are not quite as powerful as, and cannot perform the large number of engineering functions of, either the mainframe-based system or the minicomputer-based system. However, because of intense competition, the 16-bit and 32-bit systems are being produced by the major microcomputer manufacturers. These systems are desk-top models that can perform simulation, design, and drafting functions and are providing an impact in education and industry because of their low cost and increasing high technology capability. Various software programs for CAD are in existence and three-dimensional packages are available on the more powerful units. Because of the capability of the most recent systems, more and more software programs are becoming available almost daily.

16.6.1 The APPLE II GS System

The APPLE II GS system, developed by Apple Computer, utilizes an advanced 16-bit processor which runs the existing APPLE IIc and IIe software as much as 2.8 times faster than previous models. With the addition of a memory-expansion card, the APPLE II GS system takes advantage of graphics-based applications developed especially for this unit. The video capabilities of this system produce high-resolution graphics and can reproduce 4096 different colors, while its 15-voice sound generator produces natural-sounding music and speech. The APPLE II GS system is shown in Figure 16.37. Specifications, capabilities, and technical data for this system are as follows:

16.6.1.1 Central Processing Unit (CPU)

- Microprocessor: 65C816
- Clock speed: 2.8 or 1.0 megahertz; user- or software-selectable
- Address bus: 24 bits
- Data bus: 8 bits
- Address range: 16,777,216 bytes
- 16-bit registers: accumulator, two index registers, direct register, stack pointer, and program counter

16.6.1.2 Memory

- 256-kilobyte RAM, expandable by 8 megabytes
- ROM includes:
  - Control panel
  - QuickDraw II graphics routines
  - Applesoft BASIC
  - Desk Accessory Manager
  - Memory Manager
  - Free-form sound-playback capabilities
  - Apple DeskTop Bus firmware
  - SANE numerics
  - Event Manager
  - Tool Locator

FIGURE 16.37
APPLE II GS system. (Courtesy of Apple Computer, Inc.)

– Video firmware
– Drivers for built-in devices (such as mouse, serial, and disk drive ports)
– Diagnostic routines

16.6.1.3 Display Modes

Graphics

- Superhigh resolution
- 320 dots horizontally by 200 dots vertically, in up to 16 colors per line and 256 colors per screen from a palette of 4096 colors
- 640 dots horizontally by 200 dots vertically, in four or more colors per line and 128 colors per screen from a palette of 4096 colors
- Double-high-resolution (560 dots horizontally by 192 dots vertically; monochrome)

Text

- 80-column text mode (80 columns by 24 lines)

16.6.1.4 Character Sets

- 32 uppercase letters, 32 lowercase letters, 32 special characters, 32 Mouse Text characters, and 12 unique characters for each of the following international character sets: U.S., U.K., French, Danish, Spanish, Italian, German, and Swedish.

16.6.1.5 Keyboard

- Standard typewriter style
- 80 keys, plus 14-key numeric keypad
- Two Apple DeskTop Bus connectors

16.6.1.6 Interfaces

Expansion slots

- One multipurpose RAM/ROM memory-expansion slot
- Seven general-purpose input/output slots for peripheral-control cards, all fully buffered, with interrupt and DMA priority
- Disk drive port: one 19-pin D-style connector

Video output

- Analog RGB, via 15-pin D-style connector
- Composite color, via RCA phono connector

16.6.1.7 Operating Systems

- ProDOS 8 and ProDOS 16
- Pascal
- DOS 3.3
- CP/M (with appropriate coprocessor card)

16.6.2 The IBM Personal Computer AT

This unit, produced by the International Business Machine Corporation, is a powerful **personal computer (PC)**, and is an example of the direction in which educational institutions and industry are moving with respect to computer hardware. The enhanced capability of this system provides a high-speed microprocessor that enables the system to manage data two to three times faster than previous IBM PC's. It has a diskette drive that can store more than a million characters of information on a single diskette—about 600 pages of double-spaced, typed information—or a fixed disk drive capable of storing more than 20 million characters of information, which is equivalent to 10,000 pages. The system hardware is displayed in Figure 16.38.

16.6.2.1 High-speed Processor

The IBM Personal Computer AT contains a 16/24-bit Intel 80286 microprocessor which allows fast, direct access to as much as 3 MB (3,072,000 characters) of information. As a result, it can process data fast, get quicker results, and work effectively and efficiently.

16.6.2.2 Memory

The base AT model is available with 256 KB (262,144 characters) standard, which can be expanded to 512 KB with a plug-in memory module. Additional memory-expansion option cards permit memory in either the standard or enhanced model to be increased to 3 MB.

16.6.2.3 Data Storage

The AT system provides data storage in two forms—on a **diskette** or on a **fixed disk**. A slimline 1.2-MB diskette drive is standard on both IBM PC AT models. It can store a total of 1,288,800 characters of information—the equivalent of about 600 pages—on a single diskette when used with the IBM Disk Operating System (DOS) revision 3.00. The enhanced model also contains a high-speed, high-capacity fixed disk drive that provides up to 20 MB of information storage (20,480,000 characters, or 10,000 pages). As a result, dozens of programs and data files on fixed disk may be available for immediate retrieval, when needed.

16.6.2.4 Enhanced Keyboard

This keyboard is an enhanced version of the regular IBM PC keyboard. It includes the standard typewriter layout, a separate ten-key calculator key pad, and ten function keys for frequently used program com-

(a)

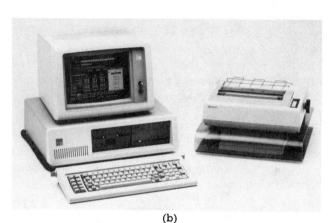

(b)

(c)

FIGURE 16.38
IBM microcomputer system: (a) PC; (b) PC XT; (c) PC AT. (Courtesy of International Business Machines Corporation.)

mands. These features are illustrated in Figure 16.39.

16.6.2.5 Display

The IBM Monochrome Display permits the user to view up to 25 lines of information with 80 characters

FIGURE 16.39
IBM PC enhanced keyboard. (Courtesy of International Business Machines Corporation.)

per line, on an 11½-inch screen (diagonal measurement). The Color Display offers a choice of 16 colors for text and eight for graphics (two sets of four colors per set) on a 12½-inch screen.

16.6.2.6 Printer

There are two dot-matrix printers available for the IBM Personal Computer AT. The graphics printer produces text and graphics at speeds of up to 80 CPS (characters per second). The color printer allows the user to produce text, charts, technical graphics, and artwork in up to eight colors at four printing speeds, from 35 CPS to 200 CPS. Figure 16.40 shows the IBM graphics printer.

FIGURE 16.40
IBM graphics printer. (Courtesy of International Business Machines Corporation.)

16.6.3 Microcomputer-based CAD Software

New software for computer-aided drafting purposes is proliferating at an unprecedented rate. Producers of existing software must keep updating and revising their programs to remain current with new products. As a result, there are many high-quality CAD software programs available to the consumer depending on need, cost, and desired features. For mechanical technical drawing purposes, the software packages identified in Table 16.1 represent areas of capability that include mechanical engineering, product design, technical illustration, tutorial options, training capability, on-line help command or menu drives, keyboard shortcuts, automatic dimensioning, automatic measuring, and technical support.

In the pages that follow, an in-depth examination of the features of four CAD software packages will be made. They include CADAPPLE, AutoCAD, VersaCAD and Micro CADAM.

16.6.3.1 The CADAPPLE 2D Software

This package was initially developed for use on an APPLE II or APPLE IIe computer configuration by T & W Systems, Inc. Its latest revision permits the user to

TABLE 16.1
Microcomputer-based CAD software.

VENDOR	PRODUCT
MICRO CADAM	MGM STATION
INNOVATIVE DATA DESIGN, INC.	MACDRAFT
AMERICAN SMALL BUSINESS COMPUTERS	PRO DESIGN II
AUTODESK	AUTOCAD
CADPAK SOFTWARE	CADPAK
CADPAK	CADPACK 2
CALCOMP SYSTEMS DIVISION	CADVANCE
EVOLUTION COMPUTING	EASYCAD
MICROCOMPUTER GRAPHICS	MGI/CADD
T&W SYSTEMS	VERSACAD
	CADAPPLE
ROBO SYSTEMS CORPORATION	ROBOCAD - PC
GENERIC SOFTWARE, INC.	GENERIC CAD

operate this software on the APPLE II GS. This package produces computer-aided drafting which includes the following convenient features:

- The program is "menu driven" and indicates to the operator that options are available.
- The operator sees on the screen exactly what the final plot will look like before it is plotted.
- The program is interactive; that is, the operator receives immediate visual feedback on all actions.
- The program has predefined "default" actions and values that enable the novice user to produce usable drawings.

Input Technology

The 2D program receives its graphics input from the digitizer (sometimes called the graphics tablet) or from the joystick. The cursor's position on the input device is represented on the screen by small crosshairs. As the drafter/designer moves the input device cursor, the crosshairs move across the screen. With the cursor, one can create various technical objects such as lines, circles, and rectangles and can graphically manipulate the entire picture or any part thereof.

Menus

This program is organized around a collection of many "menus." As with all menus, each menu has a specific purpose.

Objects

As is true with other programs on other systems, this software allows the user to create a drawing using basic building blocks called objects: basic line shapes, dimensions, points, and text.

Levels

The CADAPPLE program allows the user to manipulate up to 250 different levels.

Groups

The program also allows the user to manipulate objects in the form of **groups**. A group is an arbitrary user-defined collection of objects. The only relationship that objects in a group have to one another is that they share the same group name. Groups provide a convenient way of manipulating large components of a complex drawing.

Windows

The user may view the drawing through any user-defined viewing **window**, also called a **zoom-in, zoom-out**, or **pan** feature on other systems.

Properties

As a drawing element is created, it may be given several attributes or properties that partly determine what it will look like. An example of a property is the type of line (dashed or dotted) that will be used to plot the object.

16.6.3.2 AutoCAD Software

AutoCAD is another two-dimensional computer-aided drafting and design system that operates on low-cost microcomputers. It is produced by Autodesk, Inc. and is a multi-utility system, suitable for applications in technical drawing areas. It follows instructions to help produce a drawing quickly, offers features that allow for correction of drawing errors, and makes even large revisions without redoing an entire drawing. A completed AutoCAD drawing looks identical to the same drawing carefully prepared by hand, except that it is usually more accurate.

Equipment Requirements

The AutoCAD system is a very flexible software package and currently operates on a very high percentage of current hardware systems. In addition, AutoCAD supports all the major producers of input and output devices.

General Operation

A graphics monitor is used to display the drawing. Everything done to the drawing appears on the monitor so the user can watch each step of the progress as on other systems.

AutoCAD provides a set of entities for use in constructing the drawing. An **entity** is a drawing element also referred to as an **object** or a **primitive**. Commands are entered to tell AutoCAD which entity to draw. Commands may be typed on the keyboard, selected from the screen menu using a pointing device, or entered with the push of a button from a menu on a digitizing tablet. The program provides commands that allow modification of the drawing; entities can be erased, moved, or copied to form repeated patterns.

AutoCAD Features

- **Main menu** AutoCAD operates on two levels to reduce both the work necessary for generating a drawing and the time needed to learn the system. At the outer level, AutoCAD provides a menu-driven interface (the "main menu") that allows the user to initiate various tasks, such as creating new drawings, modifying stored drawings, producing plots, and so on. At the inner level, the menu provides access to various parts of AutoCAD, such as the Interactive Drawing Editor and the plotter interface.

- **Interactive Drawing Editor** The Drawing Editor displays the intended picture and provides commands to create, modify, view, and plot drawings.

- **Screen menu** A menu can be displayed on the graphics monitor while the drawing editor is active. This menu allows command entry simply by pointing to the command on the display screen.

- **Drawing insertion** This feature allows the operator to insert an existing AutoCAD drawing (stored on disk) into the drawing currently being created.

- **Layers and colors** Part of the drawing may be assigned to any of 127 different layers. A color may be associated with each layer. The color is simply identified by a coded number from 1 to 127.

- **Graphic input (pointing devices)** A pointing device may be used for quick command entry and to locate points in the drawing.

- **Point and command entry flexibility** One can enter commands and specify points in the drawing in a variety of ways. From the keyboard, points can be specified by typing in absolute coordinates or coordinates relative to the last point specified, or by keyboard pointing, in which keyboard control keys are used to move crosshairs around on the graphics monitor so the user can visually position to the desired point to begin a drawing.

- **Tablet menu** A menu of AutoCAD commands can be placed on the digitizing tablet, permitting entry of a command simply by pointing to it with the stylus and pushing a button.

- **Database storage** All information about the drawing, including the size and position of every element, the size of the drawing itself, and its display characteristics, is automatically updated with each command and stored on disk.

- **Plotting** A hard copy of the drawing can be plotted at any stage of its development. Typical hard-copy plotters for microcomputer-based systems are shown in Figure 16.41.

16.6.3.3 VersaCAD

The most recent generation of VersaCAD is called VersaCAD Advanced 3D Enhancement and was created for the new and advanced microcomputers by T & W Systems, Inc. It possesses advanced drawing capabilities and additional productivity benefits of lists of materials and database extraction. Working under an MS-DOS operating system, it supports hardware such as the IBM PC AT and XT, Tandy 1200 and 2000, AT&T 6300, TI Professional, and the 32-bit Hewlett-Packard Series 200 computers.

This upgraded version of VersaCAD 3D provides keyboard or digitizer input and allows the op-

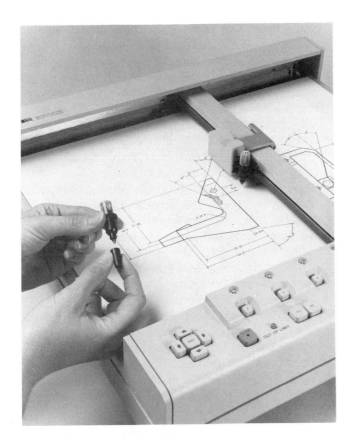

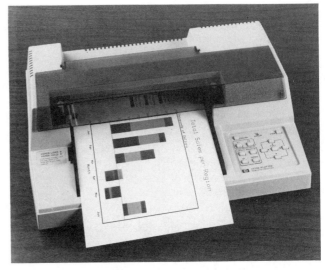

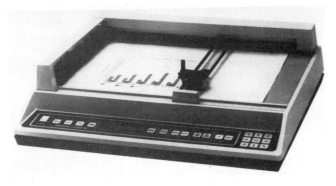

FIGURE 16.41
Hard-copy plotters for microcomputer systems. (Top left: 9872 C Plotter, courtesy of Hewlett-Packard. Top right: 7470 A Plotter, courtesy of Hewlett-Packard. Bottom left: DMP-29 Plotter, courtesy of Houston Instrument. Bottom right: DMP-40 Plotter, courtesy of Houston Instrument.)

erator to change eyepoint or center of interest with a function key and to save and retrieve up to eight views as work is being accomplished. The program operates up to three times faster than earlier releases and provides:

■ A variety of primitives, including boxes, spheres, cones, cylinders, and portions of these objects (e.g., one-half, one-quarter, and one-eighth spheres).

■ Editing features such as copy, move, or rotate along any axis, unproportional scaling, and changing of object properties.
■ Full grouping features, including building of groups by color, level number, fence, and so on.
■ Three-axis screen cursor, which always displays the true X, Y, and Z axis orientation.
■ Automatic preview of selected eyepoint and center of interest coordinates before display.
■ Display and drawing file output can be selected in either the isometric or perspective mode. 3D visu-

alization enables the operator to draw ortho-graphic views and then automatically convert them into isometric views.

- A drawing completed on the 3D enhancement can be stored as a 2D drawing file and can be retrieved later.
- Cut-and-paste capabilities speed up the editing of drawings, allowing the operator to select objects in a particular area of the drawing and then delete them or "paste" them somewhere else to be edited, scaled, rotated, or saved as new library symbols.

16.6.3.4 Micro CADAM

Micro CADAM, developed for use on the IBM Personal Computer AT by CADAM, Incorporated, a subsidiary of the Lockheed Corporation, operates on a stand-alone basis or may be used to expand versatility within a professional CADAM or a mainframe CADAM environment. It is a comprehensive software system designed to support computerized technical design/drafting and manufacturing (CAD/CAM). Its capability involves several comprehensive CAD/CAM solutions which are pictorially identified in Figure 16.42.

Micro CADAM offers the tools needed for design and drafting productivity in a wide variety of fields, including:

- **Automatic isometrics** Geometric features of any orthographic view can easily be projected onto another defined plane, generating an alternative view, such as in isometric projection.
- **Mathematical accuracy** This is a feature that permits very accurate geometric modeling. For example, if a circle is drawn, when it is projected into the isometric view it becomes an accurate representation of the original circle.
- **Dimensioning** Dimensions on drawings are more than lines and numerals. They are complete, self-contained entities.
- **Geometry grouping** Any geometry and related text may be grouped together and erased, moved, modified, or stored for later use.
- **Multiple colors and lines** Instead of segregating colors and line types on separate drawings or layers, the software allows the operator to combine them on a single drawing, without complex manipulation.
- **Detail drawings** As primary drawings are being developed, up to 63 significant features or details of that drawing can be stored for repeated or later use, or can be modified.
- **Automatic placement** Appropriate locations and spacing of features such as holes, screws, and slots can be found automatically.

In addition, project management tools such as text, nomenclature, specifications, prices, and other kinds of information may be attached to any drawing element. It can be used to generate bills of materials and similar documentation.

16.7 SUMMARY

Computer-aided drafting (CAD) has had a tremendous positive impact on the way business is conducted in engineering departments, particularly in regard to the design and drafting functions. The potential for computers in the field of engineering is overwhelming. The next generation of computers may very well be optical computers which operate a thousand times faster than anything currently in production—systems that duplicate human brain functions. One of the greatest advances has been in the field of drafting technology, especially in the production of mechanical diagrams and drawings. Productivity has increased and, at the same time, cleaner, neater, and more accurate drawings are being produced.

Equipment ranging from stand-alone mainframe systems to minicomputer and microcomputer hardware and software is available. We have examined several mainframe, mini, and micro two- and three-dimensional drafting systems. These systems are generally within the financial reach of most edu-

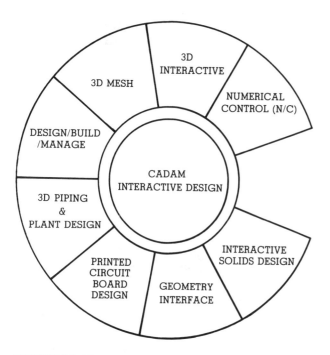

FIGURE 16.42
Comprehensive CAD/CAM solutions.

cational institutions and can be valuable tools for student and staff development, instruction, and training. We discussed, in detail, representative examples of CAD software packages. New software packages are rapidly being developed so that additional hardware and software capabilities continue to proliferate. Practice exercises for this section are presented on pages 16A through 16R in the workbook.

16.8 REVIEW EXERCISES

1. Why is a CAD system valuable?

2. Lay out freehand a block diagram of the basic concept of a CAD system.

3. Define CAD "hardware."

4. Define CAD "software."

5. What does the term "interactive" signify?

6. What three types of computer systems are available for CAD work?

7. What is a "menu tablet"?

8. What piece of hardware in a CAD system produces a "hard copy"?

9. What is the purpose of an alphanumeric keyboard?

10. What is a "floppy disk"?

11. What is the function of a cursor?

12. What is a "CPU"?

13. What is an "input device"? Give an example.

14. What is an "output device"? Give an example.

15. What kind of device is a "modem"? How is it used?

16. List six different microcomputer-based CAD software packages.

    a.

    b.

    c.

    d.

    e.

    f.

17. What is the difference between a line printer and a plotter?

18. Which CAD system is the most powerful and the most expensive?

APPENDIX A

HOLE/DRILL SIZE CHART

DRILL SIZE	DECIMAL DIA.	DRILL SIZE	DECIMAL DIA.	DRILL SIZE	DECIMAL DIA.	DRILL SIZE	DECIMAL DIA.	DRILL SIZE	DECIMAL DIA.
80	.0135	50	.0700	22	.1570	17/64	.2656	1/2	.5000
79	.0145	49	.0730	21	.1590	H	.2660	33/64	.5156
1/64	.0156	48	.0760	20	.1610	I	.2720	17/32	.5312
78	.0160	5/64	.0781	19	.1660	J	.2770	35/64	.5469
77	.0180	47	.0785	18	.1695	K	.2811	9/16	.5625
76	.0200	46	.0810	11/64	.1719	9/32	.2812	37/64	.5781
75	.0210	45	.0820	17	.1730	L	.2900	19/32	.5937
74	.0225	44	.0860	16	.1770	M	.2950	39/64	.6094
73	.0240	43	.0890	15	.1800	19/64	.2968	5/8	.6250
72	.0250	42	.0935	14	.1820	N	.3020	41/64	.6406
71	.0260	3/32	.0937	13	.1850	5/16	.3125	21/32	.6562
70	.0280	41	.0960	3/16	.1875	O	.3160	43/64	.6719
69	.0292	40	.0980	12	.1890	P	.3230	11/16	.6875
68	.0310	39	.0995	11	.1910	21/64	.3281	45/64	.7031
1/32	.0313	38	.1015	10	.1935	Q	.3320	23/32	.7187
67	.0320	37	.1040	9	.1960	R	.3390	47/64	.7344
66	.0330	36	.1065	8	.1990	11/32	.3437	3/4	.7500
65	.0350	7/64	.1093	7	.2010	S	.3480	49/64	.7656
64	.0360	35	.1100	13/64	.2031	T	.3580	25/32	.7812
63	.0370	34	.1110	6	.2040	23/64	.3594	51/64	.7969
62	.0380	33	.1130	5	.2055	U	.3680	13/16	.8125
61	.0390	32	.1160	4	.2090	3/8	.3750	53/64	.8281
60	.0400	31	.1200	3	.2130	V	.3770	27/32	.8437
59	.0410	1/8	.1250	7/32	.2187	W	.3860	55/64	.8594
58	.0420	30	.1285	2	.2210	25/64	.3906	7/8	.8750
57	.0430	29	.1360	1	.2280	X	.3970	57/64	.8906
56	.0465	28	.1405	A	.2340	Y	.4040	29/32	.9062
3/64	.0469	9/64	.1406	15/64	.2344	13/32	.4062	59/64	.9219
55	.0520	27	.1440	B	.2380	Z	.4130	15/16	.9375
54	.0550	26	.1470	C	.2420	27/64	.4219	61/64	.9531
53	.0595	25	.1495	D	.2460	7/16	.4375	31/32	.9687
1/16	.0625	24	.1520	1/4	.2500	29/64	.4531	63/64	.9844
52	.0635	23	.1540	F	.2570	15/32	.4687	1	1.000
51	.0670	5/32	1562	G	.2610	31/64	.4843		

APPENDIX A

INCHES/METRIC DECIMAL EQUIVALENTS

FRACTION			DECIMAL		FRACTION			DECIMAL	
			INCH	METRIC (mm)				INCH	METRIC (mm)
		1/64	.015625	0.3969			33/64	.515625	13.0969
	1/32		.03125	0.7938		17/32		.53125	13.4938
		3/64	.046875	1.1906			35/64	.546875	13.8906
1/16			.0625	1.5875	9/16			.5625	14.2875
		5/64	.078125	1.9844			37/64	.578125	14.6844
	3/32		.09375	2.3813		19/32		.59375	15.0813
		7/64	.109375	2.7781			39/64	.609375	15.4781
1/8			.1250	3.1750	5/8			.6250	15.8750
		9/64	.140625	3.5719			41/64	.640625	16.2719
	5/32		.15625	3.9688		21/32		.65625	16.6688
		11/64	.171875	4.3656			43/64	.671875	17.0656
3/16			.1875	4.7625	11/16			.6875	17.4625
		13/64	.203125	5.1594			45/64	.703125	17.8594
	7/32		.21875	5.5563		23/32		.71875	18.2563
		15/64	.234375	5.9531			47/64	.734375	18.6531
1/4			.250	6.3500	3/4			.750	19.0500
		17/64	.265625	6.7469			49/64	.765625	19.4469
	9/32		.28125	7.1438		25/32		.78125	19.8438
		19/64	.296875	7.5406			51/64	.796875	20.2406
5/16			.3125	7.9375	13/16			.8125	20.6375
		21/64	.328125	8.3384			53/64	.828125	21.0344
	11/32		.34375	8.7313		27/32		.84375	21.4313
		23/64	.359375	9.1281			55/64	.859375	21.8281
3/8			.3750	9.5250	7/8			.8750	22.2250
		25/64	.390625	9.9219			57/64	.890625	22.6219
	13/32		.40625	10.3188		29/32		.90625	23.0188
		27/64	.421875	10.7156			59/64	.921875	23.4156
7/16			.4375	11.1125	15/16			.9375	23.8125
		29/64	.453125	11.5094			61/64	.953125	24.2094
	15/32		.46875	11.9063		31/32		.96875	24.6063
		31/64	.484375	12.3031			63/64	.984375	25.0031
1/2			.500	12.7000	1			1.000	25.4000

APPENDIX B

TAP DRILL SIZES FOR MACHINE SCREWS

SIZE OF SCREW		NO. OF THREADS PER INCH	TAP DRILLS	
NO. OR DIAMETER	DECIMAL EQUIVALENT		DRILL SIZE	DECIMAL EQUIVALENT
0	.060	80	3/64	.0469
1	.073	64	53	.0595
		72	53	.0595
2	.086	56	50	.0700
		64	50	.0700
3	.099	48	47	.0785
		56	45	.0820
4	.112	40	43	.0890
		48	42	.0935
5	.125	40	38	.1015
		44	37	.104
6	.138	32	36	.1065
		40	33	.1130
8	.164	32	29	.1360
		36	29	.1360
10	.190	24	25	.1495
		32	21	.1590
12	.216	24	16	.1770
		28	14	.1820
1/4	.250	20	7	.2010
		28	3	.213
5/16	.3125	18	F	.2570
		24	I	.2720
3/8	.375	16	5/16	.3125
		24	Q	.3320
7/16	.4375	14	U	.3680
		20	25/64	.3906
1/2	.500	13	27/64	.4219
		20	29/64	.4531

APPENDIX C

CLEARANCE HOLE SIZES FOR MACHINE SCREWS

CLEARANCE HOLE DRILLS					
SIZE OF SCREW		CLOSE FIT		FREE FIT	
NO. OR DIAMETER	DECIMAL EQUIVALENT	DRILL SIZE	DECIMAL EQUIVALENT	DRILL SIZE	DECIMAL EQUIVALENT
0	.060	52	.0635	50	.0700
1	.073	48	.0760	46	.0810
2	.086	43	.0890	41	.0960
3	.099	37	.1040	35	.1100
4	.112	32	.1160	30	.1285
5	.125	30	.1285	29	.1360
6	.138	27	.1440	25	.1495
8	.164	18	.1695	16	.1770
10	.190	9	.1960	7	.2010
12	.216	2	.2210	1	.228
¼	.250	F	.2570	H	.2660
⁵⁄₁₆	.3125	P	.3230	Q	.3320
⅜	.375	W	.3860	X	.3970
⁷⁄₁₆	.4375	²⁹⁄₆₄	.4531	¹⁵⁄₃₂	.4687
½	.500	³³⁄₆₄	.5156	¹⁷⁄₃₂	.5312

APPENDIX D

SCREWS

SCREW SIZE NO.	DIA	PAN HEAD A	PAN HEAD H	FLAT HEAD A	FLAT HEAD H	SOCKET HEAD A	SOCKET HEAD H	HEX HEAD MACHINE A	H	P	A	H	P
2	.086	.167	.062	.172	.051	.140	.086	.125	.050	.144			
4	.112	.219	.080	.225	.067	.183	.112	.187	.060	.216			
6	.138	.270	.097	.279	.083	.226	.138	.250	.080	.288			
8	.164	.322	.115	.332	.100	.270	.164	.250	.110	.288			
10	.190	.373	.133	.385	.116	.313	.190	.312	.120	.360			
1/4	.250	.492	.175	.507	.153	.375	.250				.437	.163	.505
5/16	.312	.615	.218	.635	.191	.469	.313				.500	.211	.577
3/8	.375	.740	.261	.762	.230	.563	.375				.562	.243	.650
7/16	.437	.863	.247	.812	.223	.656	.438				.625	.316	.722
1/2	.500	.987	.281	.875	.223	.750	.500				.750	.364	.866
9/16	.562	1.04	.315	1.00	.260						.812	.371	.938
5/8	.625	1.17	.350	1.12	.298	.938	.625				.938	.444	1.08

APPENDIX E

WASHERS

	LOCK WASHER MEDIUM		FLAT WASHER		INTERNAL TOOTH LOCKWASHER		EXTERNAL TOOTH LOCKWASHER	
SIZE	OD	T	OD	T	OD	T	OD	T
2	.175	.026	.260	.025	.200	.015		
4	.212	.031	.260	.028	.270	.019	.260	.019
6	.253	.037	.385	.065	.295	.021	.320	.022
8	.296	.046	.385	.065	.340	.023	.381	.023
10	.337	.053	.448	.065	.381	.025	.410	.025
$1/4$	.493	.072	.635	.080	.478	.028	.510	.028
$5/16$	.591	.088	.697	.080	.610	.034	.610	.034
$3/8$	.688	.104	.822	.080	.692	.040	.694	.040
$7/16$	.779	.109	.922	.080	.789	.040	.760	.040
$1/2$	.873	.125	1.06	.121	.900	.045	.900	.045
$9/16$	.971	.141	1.15	.121	.985	.045	.985	.045
$5/8$	1.07	.156	1.31	.121	1.07	.050	1.07	.050

APPENDIX F

NUTS

	SQUARE NUT				HEX NUT		
SIZE	DIA	F	G	H	A	T	P
2					.187	.066	.217
4					.250	.098	.289
6					.312	.114	.361
8					.343	.130	.397
10					.375	.130	.433
¼	.250	.438	.619	.235	.437	.226	.505
5⁄16	.312	.562	.795	.283	.500	.273	.577
3⁄8	.375	.625	.884	.346	.562	.337	.650
7⁄16	.437	.750	1.06	.394	.688	.385	.794
½	.500	.812	1.14	.458	.750	.448	.866
9⁄16					.875	.496	1.01
5⁄8	.625	1.00	1.41	.569	.938	.559	1.08

APPENDIX G

MILITARY APPROVED ABBREVIATIONS FOR COMMONLY USED WORDS AND WORD COMBINATIONS

GENERAL RULES

a. Abbreviations of word combinations shall not be separated for single use.

b. Single abbreviations may be combined when necessary.

c. Spaces between word combination abbreviations may be filled with a hyphen (-) for clarity.

d. The same abbreviations shall be used for all tenses, the possessive case, singular, and plural forms of a given word.

WORD	ABBREVIATION	WORD	ABBREVIATION
Absolute	ABS	April	APR
Accessory	ACCESS.	Arc Weld	ARC W
Accumulate	ACCUM	Armature	ARM.
Actual	ACT.	Assemble	ASSEM
Adapter	ADPT	Assembly	ASSY
Addendum	ADD.	Attention	ATTN
Addition	ADD.	Attenuation, Attenuator	ATTEN
Adjust	ADJ	Audio frequency	AF
After	AFT.	August	AUG
Alignment	ALIGN.	Authorize	AUTH
Allowance	ALLOW.	Automatic frequency control	AFC
Alloy	ALY	Automatic gain control	AGC
Alteration	ALT	Automatic volume control	AVC
Alternate	ALT	Auxiliary	AUX
Alternating current	AC	Auxiliary power unit	APU
Ambient	AMB	Average	AVG
American wire gage	AWG	Avoirdupois	AVDP
Amount	AMT	Azimuth	AZ
Ampere	AMP	Balance	BAL
Ampere (combination form)	A	Ball bearing	BB
Ampere hour	AMP HR	Bandpass	BP
Amplifier	AMPL	Bandwidth	B
Amplitude modulation	AM.	Base line	BL
And	&	Basic	BSC
Annunciator	ANN	Battery (electrical)	BAT.
Anodize	ANOD	Bearing	BRG.
Antenna	ANT.	Beat-frequency (adj)	B F
Application	APPL	Beat-frequency oscillator	BFO
Approved	APPD	Beryllium	BE.
Approximate	APPROX	Between	BET.

APPENDIX G

MILITARY APPROVED ABBREVIATIONS

WORD	ABBREVIATION	WORD	ABBREVIATION
Between centers	BC	Change order	CO
Bill of Material	B/M	Chassis	CHAS
Binding	BIND.	Chromium	Cr
Black	BLK	Circuit	CKT
Blank	BLK	Circuit breaker	CKT BKR
Blower	BLO	Circular	CIR
Blue	BLU	Circular pitch	CP
Board	BD	Circumference	CIRC
Bolt circle	BC	Class	CL
Both faces	BF	Clearance	CL
Both sides	BS	Clockwise	CW
Bottom face	BF	Coaxial	COAX.
Bracket	BRKT	Coefficient	COEF
Brass	BRS	Cold-drawn steel	CDS
Brazing	BRZG	Cold-rolled	CR
Bridge	BRDG	Cold-rolled steel	CRS
Brinell hardness	BH	Collector	COLL
Brinell hardness number	BHN	Color code	CC
British thermal unit	BTU	Commercial	COML
Bronze	BRZ	Common	COM
Brown	BRN	Company	CO
Brown and Sharpe (gage)	B&S	Composition	COMP
Building	BLDG	Concentric	CONC
Burnish	BHN	Condition	COND
Bushing	BUSH.	Conductor	COND
Buzzer	BUZ	Connector	CONN
Bypass	BYP	Contact	CONT
Cadmium	CAD.	Continue	CONT
Calibrate	CAL	Continuous wave	CW
Camber	CAM.	Control	CONT
Capacitor	CAP.	Copper	COP
Capacity	CAP.	Corporation	CORP
Carbon	C	Corrosion	CORR
Carbon Steel	CS	Corrosion-resistant	CRE
Carton	CTN	Corrosion-resistant steel	CRES
Castellate	CTL	Cotangent	COT.
Casting	CSTG	Counterbore	CBORE
Cast Iron	CI	Counterclockwise	CCW
Cathode-ray tube	CRT	Counterdrill	CDRILL
Center	CTR	Countersink	CSK
Center line	CL or ℄	Countersink other side	CSKO
Center of gravity	CG	Cross section	XSECT
Center Tap	CT	Crystal	XTAL
Center to Center	C TO C	Cubic	CU
Centigrade	C	Cubic centimeter	CC
Centimeter	CM	Cubic feet	CU FT
Ceramic	CER	Cubic feet per minute	CFM
Chamfer	CHAM	Cubic feet per second	CFS
Change	CHG	Cubic inch	CU IN.
Change notice	CN	Cubic meter	CU M

APPENDIX G

MILITARY APPROVED ABBREVIATIONS

WORD	ABBREVIATION	WORD	ABBREVIATION
Cubic millimeter	CU MM	Engineer	ENGR
Current	CUR.	Engineering	ENGRG
Cycle	CY	Equipment	EQUIP.
Cycles per minute	CPM	Equivalent	EQUIV
Cycles per second	CPS	Estimate	EST
Datum	D	Et cetera	ETC
Decalcomania	DECAL	Example	EX
December	DEC	Exclusive	EXCL
Decibel	DB	Extension	EXT
Decimal	DEC	External	EXT
Deep-drawn	DD	Fahrenheit	F
Degree	DEG or °	Farad	F
Delay	DLY	Far side	FS
Delay line	DL	Fastener	FASTNR
Depth	D	February	FEB
Detail	DET	Federal	FED.
Deviation	DEV	Federal Specification	FS
Diagram	DIAG	Federal stock number	FSN
Diameter	DIA	Figure	FIG.
Diameter bolt circle	DBC	Filament	FIL
Diametral pitch	DP	Filament center tap	FCT
Dimension	DIM.	Fillister head	FIL H
Direct current	DC	Finish	FIN.
Direct-current volts	VDC	Finish all over	FAO
Direct-current working volts	VDCW	Fixed	FXD
Disconnect	DISC.	Flange	FLG
Division	DIV	Flat head	FH
Door	DR	Foot	FT or '
Double pole	DP	Foot-pound	FT LB
Double-pole double throw	DPDT	For example	EG
Double-pole single throw	DPST	Four-pole	4P
Double throw	DT	Frequency	FREQ
Down	DN	Frequency modulation	FM
Drawing	DWG	Front	FR
Drawing list	DL	Gage	GA
Drill	DR	Gallon	GAL
Drill rod	DR	Galvanize	GALV
Drive	DR	Half-hard	½ H
Drive fit	DF	Half-round	½ RD
Duplicate	DUP	Hard	H
Each	EA	Harden	HDN
Eccentric	ECC	Hardware	HDW
Effective	EFF	Head	HD
Electric	ELEC	Headless	HDLS
Electrolytic	ELECT.	Heater	HTR
Element	ELEM	Height	HGT
Eliminate	ELIM	Henry (electrical)	H
Elongation	ELONG	Hertz	HZ
Emergency	EMER	Hexagon	HEX.
End to end	E to E	Hexagonal head	HEX HD

APPENDIX G

MILITARY APPROVED ABBREVIATIONS

WORD	ABBREVIATION	WORD	ABBREVIATION
High frequency	HF	Machine screw	MS
High frequency oscillator	HFO	Maintenance	MAINT
Horizon, Horizontal	HORIZ	Major	MAJ
Horizontal center line	HCL	Manual	MAN
Horsepower	HP	Manufacture	MFR
Hot-rolled steel	HRS	Manufactured	MFD
Hour	HR	Manufacturing	MFG
Identification	IDENT	March	MAR.
Impeller	IMP.	Master oscillator	MO
Impulse	IMP.	Material	MATL
Inch	IN. or "	Maximum	MAX
Inches per second	IPS	Mechanical	MECH
Inch-pound	IN LB	Megacycle	MC
Include	INCL	Megacycle per second	MCS
Inclusive	INCL	Megohm	MEGO
Incorporated	INC	Meter	M
Indicate	IND	Microfarad	UF or uF
Indicator	IND	Microhenry	UH or uH
Information	INFO	Microhm	UOHM or
Inside diameter	ID		uOHM
Inside radius	IR	Microinch	UIN or uIN
Insulation, Insulator	INS	Micro micro (10^{-12})	UU or uu
Intermediate frequency	IF	Micromicrohenry	UUH or uuH
Internal	INT	Micromicrofarad	UUF or uuF
Irregular	IRREG	Micromicron	UU or uu
Issue	ISS	Micron (.001 millimeter)	U or u
Jack	J	Milliampere	MA
January	JAN	Millihenry	MH
July	JUL	Millimeter	MM
Junction	JCT	Millivolt	MV
Junction box	JB	Milliwatt	MW
June	JUN	Minimum	MIN
Keyway	KWY	Minor	MIN
Kilocycle	KC	Miscellaneous	MISC
Kilocycles per second	KC/S	Model (for general use)	MOD
Left	L	Modification	MOD
Left hand	LH	Modify	MOD
Length	LG	Modulator	MOD
Length over-all	LOA	Mount	MT
Light	LT	Mounting	MTG
Linear	LIN	National coarse (thread)	NC
Liquid	LIQ	National extra fine (thread)	NEF
Load limiting resistor	LLR	National fine (thread)	NF
Local oscillator	LO	National special (thread)	NS
Lock washer	LK WASH.	Nickel	NI
Long	LG	Nominal	NOM
Loudspeaker	LS	Normal	NORM.
Low frequency	LF	Normally closed	NC
Low-frequency oscillator	LFO	Normally open	NO.
Low pass	LP	Not applicable	NA

APPENDIX G

MILITARY APPROVED ABBREVIATIONS

WORD	ABBREVIATION	WORD	ABBREVIATION
Not to scale	NTS	Pound-foot	LB FT
November	NOV	Pounds per square inch	PSI
Number	NO.	Power	PWR
Nylon	N	Power amplifier	PA
Obsolete	OBS	Power supply	PWR SUP
October	OCT	Preamplifier	PREAMP
Ohm (for use only on diagrams)	Ω	Preferred	PFD
On center	OC	Preliminary	PRELIM
Opposite	OPP	Prepare	PREP
Optional	OPT	Primary	PRIM.
Orange	ORN	Project	PROJ
Origin, Original	ORIG	Quantity	QTY
Oscillator	OSC	Quarter-hard	¼ H
Ounce	OZ	Quartz	QTZ
Ounce-inch	OZ IN	Radio frequency	RF
Output	OUPT	Radius	R or RAD.
Outside diameter	OD	Rate	RT
Outside radius	OR.	Reactor	REAC
Oval head	OV HD	Received	RECD
Page	P	Receiver	RCVR
Pair	PR	Receptable	RECP
Panel	PNL	Rectifier	RECT
Paragraph	PARA	Reference	REF
Parallel	PAR.	Reference line	REF L
Part	PT	Relay	REL
Part number	PN	Release	REL
Passivate	PASS.	Relief	REL
Per	/	Remove	REM
Per centum	PCT	Required	REQD
Perpendicular	PERP	Resistor	RES.
Phase	PH	Revision	REV
Phenolic	PHEN	Rheostat	RHEO
Phillips head	PHL H	Right	R
Phosphor bronze	PH BRZ	Right hand	RH
Piece	PC	Rivet	RIV
Pitch	P	Rockwell hardness	RH
Pitch circle	PC	Rotary	ROT.
Pitch diameter	PD	Round	RD
Plastic	PLSTC	Roundhead	RH
Plate	PL	Schedule	SCH
Plate (electron tube)	P	Schematic	SCHEM
Plug	PL	Screw	SCR
Plus or minus	±	Seamless	SMLS
Point	PT	Seamless steel tubing	SSTU
Point of intersection	PI	Section	SECT.
Point of tangency	PT	Selector	SEL
Pole	P	September	SEP
Position	POS	Serial	SER
Positive	POS	Servo	SVO
Pound	LB	Set screw	SS

APPENDIX G

MILITARY APPROVED ABBREVIATIONS

WORD	ABBREVIATION	WORD	ABBREVIATION
Sheet	SH	Three-pole	3P
Shield	SHLD	Through	THRU
Shoulder	SHLD	Time delay	TD
Silver	SIL	Toggle	TGL
Similar	SIM	Tolerance	TOL
Single pole	SP	Torque	TOR
Single pole, double throw	SPDT	Transformer	XMFR
Single pole, single throw	SPST	Transistor	TSTR
Single throw	ST	Transmitter	XMTR
Slotted	SLOT.	Transmitter receiver	TR
Small	SM	Triple pole	3P
Socket	SOC	Triple throw	3T
Socket head	SCH	True position	TP
Solenoid	SOL.	Tubing	TUB.
Spacer	SPR	Twisted	TW
Speaker	SPKR	Typical	TYP
Special	SPL	Unfinished	UNFIN
Specification	SPEC	Unified coarse thread	UNC
Spectrum analyzer	SA	Unified extra fine thread	UNEF
Spherical	SPHER	Unified fine thread	UNF
Split ring	SR	Unified special thread	UNS
Spot face	SF	United States Air Force	USAF
Spot weld	SW	Upper	UP.
Spring	SPG	Used with	U/W
Square	SQ	Variable	VAR
Square foot	SQ FT	Variable frequency oscillator	VFO
Square inch	SQ IN.	Vertical	VERT.
Stainless steel	SST	Vertical center line	VCL
Standard	STD	Very-high frequency	VHF
Steel	STL	Very-low frequency	VLF
Stranded	STRD	Video	VID
Surface	SURF.	Viscosity	VIS
Switch	SW	Volt	V
Symbol	SYM	Washer	WASH.
Symmetrical	SYM	Watt	W
Tangent	TAN.	Weight	WT
Tapping	TAP.	White	WHT
Tempered	TEMP	Wire-wound	WW
Terminal	TERM.	With (abbreviate only in	
That is	i.e.	conjunction with other	
Thermal	THRM	abbreviations)	W/
Thermistor	TMTR	Without	W/O
Thick	THK	Yellow	YEL
Thread	THD	Zinc	Zn
Threads per inch	TPI		

APPENDIX H

STANDARD SCREW THREAD CHART

Nominal Size & Threads Per Inch	Series Desig.	Class	Ext. Allowance	Ext. Major Dia Max	Ext. Major Dia Min	Ext. —	Ext. Pitch Dia Max	Ext. Pitch Dia Min	Ext. Pitch Dia Tol.	Minor Dia	Class	Int. Minor Dia Min	Int. Minor Dia Max	Int. Pitch Dia Min	Int. Pitch Dia Max	Int. Pitch Dia Tol.	Int. Major Dia Min
1-64	UNC	2A	.0006	.0724	.0686	—	.0623	.0603	.0020	.0532	2B	.0561	.0623	.0629	.0655	.0026	.0730
		3A	.0000	.0730	.0692	—	.0629	.0614	.0015	.0538	3B	.0561	.0623	.0629	.0648	.0119	.0730
2-56	UNC	2A	.0006	.0854	.0813	—	.0738	.0717	.0021	.0635	2B	.0667	.0737	.0744	.0772	.0028	.0860
		3A	.0000	.0860	.0819	—	.0744	.0728	.0016	.0641	3B	.0667	.0737	.0744	.0765	.0021	.0860
3-48	UNC	2A	.0007	.0983	.0938	—	.0848	.0825	.0023	.0727	2B	.0764	.0845	.0855	.0885	.0030	.0990
		3A	.0000	.0990	.0945	—	.0855	.0838	.0017	.0734	3B	.0764	.0845	.0855	.0877	.0022	.0990
4-40	UNC	2A	.0008	.1112	.1061	—	.0950	.0925	.0025	.0805	2B	.0849	.0939	.0958	.0991	.0033	.1120
		3A	.0000	.1120	.1069	—	.0958	.0939	.0019	.0813	3B	.0849	.0939	.0958	.0982	.0024	.1120
5-40	UNC	2A	.0008	.1242	.1191	—	.1080	.1054	.0026	.0935	2B	.0979	.1062	.1088	.1121	.0033	.1250
		3A	.0000	.1250	.1199	—	.1088	.1069	.0019	.0943	3B	.0979	.1062	.1088	.1113	.0025	.1250
6-32	UNC	2A	.0008	.1372	.1312	—	.1169	.1141	.0028	.0989	2B	.104	.114	.1177	.1214	.0037	.1380
		3A	.0000	.1380	.1320	—	.1177	.1156	.0021	.0997	3B	.1040	.1140	.1177	.1204	.0027	.1380
6-40	UNF	2A	.0008	.1372	.1321	—	.1210	.1184	.0026	.1065	2B	.111	.119	.1218	.1252	.0034	.1380
		3A	.0000	.1380	.1329	—	.1218	.1198	.0020	.1073	3B	.1110	.1186	.1218	.1243	.0025	.1380
8-32	UNC	2A	.0009	.1631	.1571	—	.1428	.1399	.0029	.1248	2B	.130	.139	.1437	.1475	.0038	.1640
		3A	.0000	.1640	.1580	—	.1437	.1415	.0022	.1257	3B	.1300	.1389	.1437	.1465	.0028	.1640
8-36	UNF	2A	.0008	.1632	.1577	—	.1452	.1424	.0028	.1291	2B	.134	.142	.1460	.1496	.0036	.1640
		3A	.0000	.1640	.1585	—	.1460	.1439	.0021	.1299	3B	.1340	.1416	.1460	.1487	.0027	.1640
10-24	UNC	2A	.0010	.1890	.1818	—	.1619	.1586	.0033	.1379	2B	.145	.156	.1629	.1672	.0043	.1900
		3A	.0000	.1900	.1828	—	.1629	.1604	.0025	.1389	3B	.1450	.1555	.1629	.1661	.0032	.1900
10-32	UNF	2A	.0009	.1891	.1831	—	.1688	.1658	.0030	.1508	2B	.156	.164	.1697	.1736	.0039	.1900
		3A	.0000	.1900	.1840	—	.1697	.1674	.0023	.1517	3B	.1560	.1641	.1697	.1726	.0029	.1900
12-24	UNC	2A	.0010	.2150	.2078	—	.1879	.1845	.0034	.1639	2B	.171	.181	.1889	.1933	.0044	.2160
		3A	.0000	.2160	.2088	—	.1889	.1863	.0026	.1649	3B	.1710	.1807	.1889	.1922	.0033	.2160
12-28	UNF	2A	.0010	.2150	.2085	—	.1918	.1886	.0032	.1712	2B	.177	.186	.1928	.1970	.0042	.2160
		3A	.0000	.2160	.2095	—	.1928	.1904	.0024	.1722	3B	.1770	.1857	.1928	.1959	.0031	.2160

APPENDIX H

STANDARD SCREW THREAD CHART (CONTINUED)

NOMINAL SIZE AND THREADS PER INCH	SERIES DESIGNATION	CLASS	ALLOWANCE	MAJOR DIA LIMITS MAX	MAJOR DIA LIMITS MIN	MAJOR DIA LIMITS MIN	EXTERNAL PITCH DIA LIMITS MAX	EXTERNAL PITCH DIA LIMITS MIN	TOLERANCE	MINOR DIA	CLASS	MINOR DIA LIMITS MIN	MINOR DIA LIMITS MAX	INTERNAL PITCH DIA LIMITS MIN	INTERNAL PITCH DIA LIMITS MAX	TOLERANCE	MAJOR DIA MIN
1/4-20	UNC	1A	.0011	.2489	.2367	—	.2164	.2108	.0056	.1876	1B	.196	.207	.2175	.2248	.0073	.2500
		2A	.0011	.2489	.2408	.2367	.2164	.2127	.0037	.1876	2B	.196	.207	.2175	.2223	.0048	.2500
		3A	.0000	.2500	.2419	—	.2175	.2147	.0028	.1887	3B	.1960	.2067	.2175	.2211	.0036	.2500
1/4-28	UNF	1A	.0010	.2490	.2392	—	.2258	.2208	.0050	.2052	1B	.211	.220	.2268	.2333	.0065	.2500
		2A	.0010	.2490	.2425	—	.2258	.2225	.0033	.2052	2B	.211	.220	.2268	.2311	.0043	.2500
		3A	.0000	.2500	.2435	—	.2268	.2243	.0025	.2062	3B	.2110	.2190	.2268	.2300	.0032	.2500
5/16-18	UNC	1A	.0012	.3113	.2982	—	.2752	.2691	.0061	.2431	1B	.252	.265	.2764	.2843	.0079	.3125
		2A	.0012	.3113	.3026	.2982	.2752	.2712	.0040	.2431	2B	.252	.265	.2764	.2817	.0053	.3125
		3A	.0000	.3125	.3038	—	.2764	.2734	.0030	.2443	3B	.2520	.2630	.2764	.2803	.0039	.3125
5/16-24	UNF	1A	.0011	.3114	.3006	—	.2843	.2788	.0055	.2603	1B	.267	.277	.2854	.2925	.0071	.3125
		2A	.0011	.3114	.3042	—	.2843	.2806	.0037	.2603	2B	.267	.277	.2854	.2902	.0048	.3125
		3A	.0000	.3125	.3053	—	.2854	.2827	.0027	.2614	3B	.2670	.2754	.2854	.2890	.0036	.3125
3/8-16	UNC	1A	.0013	.3737	.3595	—	.3331	.3266	.0065	.2970	1B	.307	.321	.3344	.3429	.0085	.3750
		2A	.0013	.3737	.3643	.3595	.3331	.3287	.0044	.2970	2B	.307	.321	.3344	.3401	.0057	.3750
		3A	.0000	.3750	.3656	—	.3344	.3311	.0033	.2983	3B	.3070	.3182	.3344	.3387	.0043	.3750
3/8-24	UNF	1A	.0011	.3739	.3631	—	.3468	.3411	.0057	.3228	1B	.330	.340	.3479	.3553	.0074	.3750
		2A	.0011	.3739	.3667	—	.3468	.3430	.0038	.3228	2B	.330	.340	.3479	.3528	.0049	.3750
		3A	.0000	.3750	.3678	—	.3479	.3450	.0029	.3239	3B	.3300	.3372	.3479	.3516	.0037	.3750
7/16-14	UNC	1A	.0014	.4361	.4205	—	.3897	.3826	.0071	.3485	1B	.360	.376	.3911	.4003	.0092	.4375
		2A	.0014	.4361	.4258	.4206	.3897	.3850	.0047	.3485	2B	.360	.376	.3911	.3972	.0061	.4375
		3A	.0000	.4375	.4272	—	.3911	.3876	.0035	.3499	3B	.3600	.3717	.3911	.3957	.0046	.4375
7/16-20	UNF	1A	.0013	.4362	.4240	—	.4037	.3975	.0062	.3749	1B	.383	.395	.4050	.4131	.0081	.4375
		2A	.0013	.4362	.4281	—	.4037	.3995	.0042	.3749	2B	.383	.395	.4050	.4104	.0054	.4375
		3A	.0000	.4375	.4294	—	.4050	.4019	.0031	.3762	3B	.3830	.3916	.4050	.4091	.0041	.4375
1/2-13	UNC	1A	.0015	.4985	.4822	—	.4485	.4411	.0074	.4041	1B	.417	.434	.4500	.4597	.0097	.5000
		2A	.0015	.4985	.4876	.4822	.4485	.4435	.0050	.4041	2B	.417	.434	.4500	.4565	.0065	.5000
		3A	.0000	.5000	.4891	—	.4500	.4463	.0037	.4056	3B	.4170	.4284	.4500	.4548	.0048	.5000
1/2-20	UNF	1A	.0013	.4987	.4865	—	.4662	.4598	.0064	.4374	1B	.446	.457	.4675	.4759	.0084	.5000
		2A	.0013	.4987	.4906	—	.4662	.4619	.0043	.4374	2B	.446	.457	.4675	.4731	.0056	.5000
		3A	.0000	.5000	.4918	—	.4675	.4643	.0032	.4387	3B	.4460	.4537	.4675	.4717	.0042	.5000

APPENDIX I

PINS

SQUARE END STRAIGHT PIN

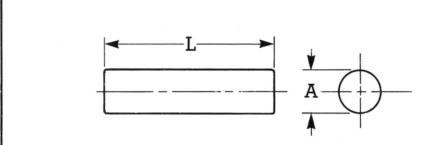

BREAK CORNERS TO .003 — .015 R OR CHAMFER

NOMINAL SIZE		PIN DIAMETER (A)	
		MAX	MIN
$\frac{1}{16}$	.062	.0625	.0605
$\frac{3}{32}$	.094	.0937	.0917
$\frac{1}{8}$	.125	.1250	.1230
$\frac{5}{32}$	.156	.1562	.1542
$\frac{3}{16}$	.188	.1875	.1855
$\frac{1}{4}$	.250	.2500	.2480
$\frac{5}{16}$	.312	.3125	.3105
$\frac{3}{8}$	.375	.3750	.3730
$\frac{7}{16}$	.438	.4375	.4335
$\frac{1}{2}$	.500	.5000	.4980
$\frac{5}{8}$	.625	.6250	.6230

(ANSI B 18.8.2 - 1978)

APPENDIX I

PINS

CLEVIS PIN

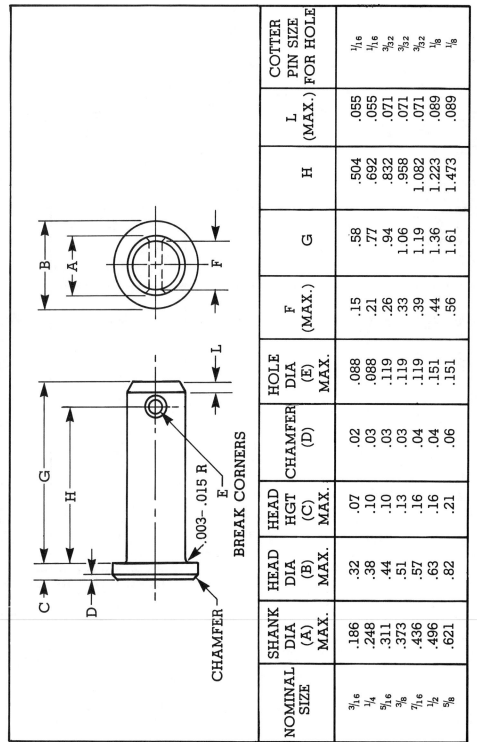

NOMINAL SIZE	SHANK DIA (A) MAX.	HEAD DIA (B) MAX.	HEAD HGT (C) MAX.	CHAMFER (D)	HOLE DIA (E) MAX.	F (MAX.)	G	H	L (MAX.)	COTTER PIN SIZE FOR HOLE
3/16	.186	.32	.07	.02	.088	.15	.58	.504	.055	1/16
1/4	.248	.38	.10	.03	.088	.21	.77	.692	.055	1/16
5/16	.311	.44	.10	.03	.119	.26	.94	.832	.071	3/32
3/8	.373	.51	.13	.03	.119	.33	1.06	.958	.071	3/32
7/16	.436	.57	.16	.04	.119	.39	1.19	1.082	.071	3/32
1/2	.496	.63	.16	.04	.151	.44	1.36	1.223	.089	1/8
5/8	.621	.82	.21	.06	.151	.56	1.61	1.473	.089	1/8

(ANSI B 18.8.1 - 1972 - R1977)

APPENDIX I

PINS

COTTER PIN - EXTENDED PRONG SQUARE CUT TYPE

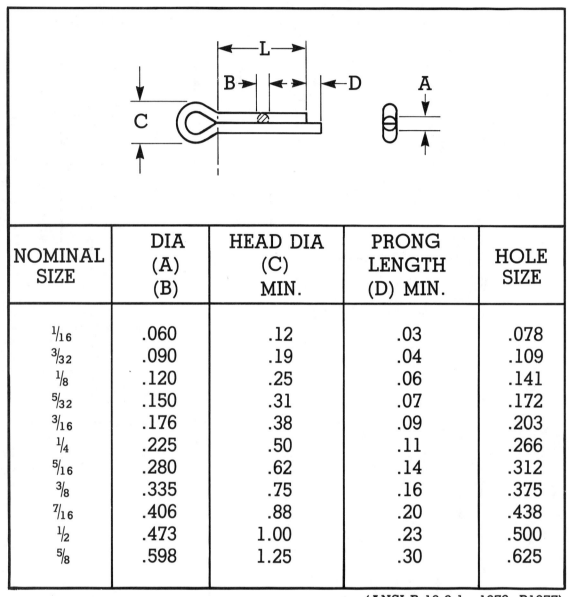

NOMINAL SIZE	DIA (A) (B)	HEAD DIA (C) MIN.	PRONG LENGTH (D) MIN.	HOLE SIZE
$\frac{1}{16}$	.060	.12	.03	.078
$\frac{3}{32}$	.090	.19	.04	.109
$\frac{1}{8}$	.120	.25	.06	.141
$\frac{5}{32}$	.150	.31	.07	.172
$\frac{3}{16}$	.176	.38	.09	.203
$\frac{1}{4}$	.225	.50	.11	.266
$\frac{5}{16}$	.280	.62	.14	.312
$\frac{3}{8}$	.335	.75	.16	.375
$\frac{7}{16}$	.406	.88	.20	.438
$\frac{1}{2}$	.473	1.00	.23	.500
$\frac{5}{8}$	.598	1.25	.30	.625

(ANSI B 18.8.1 - 1972, R1977)

APPENDIX J

BENDS - RADII IN SHEET METALS

PREFERRED BEND RADII FOR STRAIGHT BENDS IN SHEET METALS

SHEET THICKNESS	ALUMINUM						
	2024 T3	5052 0	5052 H32	5052 H34	6061 0	6061 T4	6061 T6
.020	.06	.03	.03	.03	.03	.06	.06
.025	.06	.03	.03	.03	.03	.06	.06
.032	.09	.03	.06	.06	.03	.09	.09
.040	.12	.06	.06	.06	.06	.12	.12
.050	.16	.06	.09	.09	.06	.16	.16
.063	.19	.06	.09	.09	.06	.19	.19
.080	.25	.09	.12	.12	.09	.25	.25
.090	.31	.09	.12	.12	.09	.31	.31
.100	.38	.12	.19	.19	.12	.38	.38
.125	.50	.12	.19	.19	.12	.50	.50
.160	.75	.16	.25	.25	.16	.62	.62
.190	1.00	.19	.37	.37	.19	.84	.87

SHEET THICKNESS	STEEL		PLAIN CARBON
	CORROSION RESISTANT		
	TYPES 301 302 304 (Annealed)	TYPES 301 302 304 (1/4 H)	
.020			.06
.025			.06
.032			.06
.040			.09
.050			.09
.063			.09
.080			.12
.090			.12
.100			.16
.125			.19
.160			.25
.190			.31
.020-.040	.03	.06	
.045-.070	.06	.12	
.075-.105	.09	.19	
.110-.135	.12	.25	

APPENDIX J

BENDS - 90 DEGREE DEVELOPED LENGTH

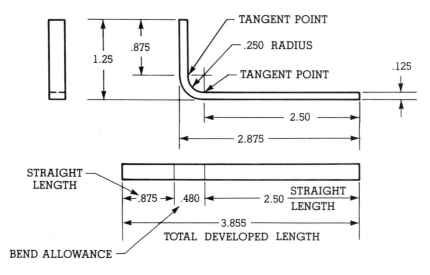

To determine the developed length of an object with a 90 degree bend.

1. Locate the tangent points at the inside of the .250 inch radius. The tangent point is where the straight length meets the radius.
2. Find the material thickness on the drawing: .125 inch.
3. Refer to the chart below where the horizontal row at the top indicates the inside radius and the left hand column indicates the material thickness. The chart shows the .250 radius and the .125 thickness meet at .480.

This dimension is called the bend allowance and is the length that must be added to the straight lengths to determine the total developed length.

BEND ALLOWANCE FOR 90° BENDS (INCH)

Radii Thickness	.031	.063	.094	.125	.156	.188	.219	.250	.281	.313	.344	.375	.438	.500
.013	.058	.108	.157	.205	.254	.304	.353	.402	.450	.501	.549	.598	.697	.794
.016	.060	.110	.159	.208	.256	.307	.355	.404	.453	.503	.552	.600	.699	.796
.020	.062	.113	.161	.210	.259	.309	.358	.406	.455	.505	.554	.603	.702	.799
.022	.064	.114	.163	.212	.260	.311	.359	.408	.457	.507	.556	.604	.703	.801
.025	.066	.116	.165	.214	.263	.313	.362	.410	.459	.509	.558	.607	.705	.803
.028	.068	.119	.167	.216	.265	.315	.364	.412	.461	.511	.560	.609	.708	.805
.032	.071	.121	.170	.218	.267	.317	.366	.415	.463	.514	.562	.611	.710	.807
.038	.075	.126	.174	.223	.272	.322	.371	.419	.468	.518	.567	.616	.715	.812
.040	.077	.127	.176	.224	.273	.323	.372	.421	.469	.520	.568	.617	.716	.813
.050		.134	.183	.232	.280	.331	.379	.428	.477	.527	.576	.624	.723	.821
.064		.144	.192	.241	.290	.340	.389	.437	.486	.536	.585	.634	.732	.830
.072			.198	.247	.296	.346	.394	.443	.492	.542	.591	.639	.738	.836
.078			.202	.251	.300	.350	.399	.447	.496	.546	.595	.644	.743	.840
.081			.204	.253	.302	.352	.401	.449	.498	.548	.598	.646	.745	.842
.091			.212	.260	.309	.359	.408	.456	.505	.555	.604	.653	.752	.849
.094			.214	.262	.311	.361	.410	.459	.507	.558	.606	.655	.754	.851
.102				.268	.317	.367	.416	.464	.513	.563	.612	.661	.760	.857
.109				.273	.321	.372	.420	.469	.518	.568	.617	.665	.764	.862
▶.125				.284	.333	.383	.432	▶.480	.529	.579	.628	.677	.776	.873
.156					.355	.405	.453	.502	.551	.601	.650	.698	.797	.895
.188						.427	.476	.525	.573	.624	.672	.721	.820	.917
.203								.535	.584	.634	.683	.731	.830	.928
.218								.546	.594	.645	.693	.742	.841	.938
.234								.557	.606	.656	.705	.753	.852	.950
.250								.568	.617	.667	.716	.764	.863	.961

APPENDIX J

GREATER THAN 90 DEGREE DEVELOPED LENGTH

BEND ALLOWANCE FOR EACH 1° OF BEND (INCH)

Radii → Thickness ↓	.031	.063	.094	.125	.156	.188	.219	.250	.281	.313	.344	.375	.438	.500
.013	.00064	.00120	.00174	.00228	.00282	.00338	.00392	.00446	.00500	.00556	.00610	.00664	.00774	.00883
.016	.00067	.00122	.00176	.00231	.00285	.00342	.00395	.00449	.00503	.00559	.00613	.00667	.00777	.00885
.020	.00069	.00125	.00179	.00233	.00287	.00343	.00397	.00452	.00506	.00561	.00616	.00670	.00780	.00888
.022	.00071	.00127	.00181	.00235	.00289	.00345	.00399	.00453	.00508	.00563	.00617	.00672	.00782	.00890
.025	.00074	.00129	.00184	.00238	.00292	.00348	.00402	.00456	.00510	.00566	.00610	.00674	.00784	.00892
.028	.00076	.00132	.00186	.00240	.00294	.00350	.00404	.00458	.00512	.00568	.00622	.00676	.00786	.00894
.032	.00079	.00134	.00189	.00243	.00297	.00353	.00407	.00461	.00515	.00571	.00625	.00679	.00789	.00897
.038	.00084	.00140	.00194	.00248	.00302	.00358	.00412	.00466	.00520	.00576	.00630	.00684	.00794	.00902
.040	.00085	.00141	.00195	.00249	.00303	.00359	.00413	.00468	.00522	.00577	.00632	.00686	.00796	.00904
.050		.00149	.00203	.00258	.00312	.00368	.00422	.00476	.00530	.00586	.00640	.00694	.00804	.00912
.064		.00160	.00214	.00268	.00322	.00378	.00432	.00486	.00540	.00596	.00650	.00704	.00814	.00922
.072			.00220	.00274	.00328	.00384	.00438	.00492	.00546	.00602	.00656	.00710	.00820	.00929
.078			.00225	.00279	.00333	.00389	.00443	.00497	.00551	.00607	.00661	.00715	.00825	.00933
.081			.00227	.00281	.00335	.00391	.00445	.00499	.00554	.00609	.00664	.00718	.00828	.00936
.091			.00235	.00289	.00343	.00399	.00453	.00507	.00561	.00617	.00671	.00725	.00835	.00944
.094			.00237	.00291	.00346	.00401	.00456	.00510	.00564	.00620	.00674	.00728	.00838	.00946
.102				.00298	.00352	.00408	.00462	.00516	.00570	.00626	.00680	.00734	.00844	.00952
.109				.00303	.00357	.00413	.00467	.00521	.00575	.00631	.00685	.00739	.00849	.00958
.125				.00316	.00370	.00426	.00480	.00534	.00588	.00644	.00698	.00752	.00862	.00970
.156					.00394	.00450	.00504	.00558	.00612	.00668	.00722	.00776	.00886	.00994
.188						.00475	.00529	.00583	.00637	.00693	.00747	.00802	.00911	.01019
.203								.00595	.00649	.00704	.00759	.00813	.00923	.01031
.218								.00606	.00660	.00716	.00770	.00824	.00934	.01042
.234								.00619	.00673	.00729	.00783	.00837	.00947	.01055
.250								.00631	.00685	.00741	.00795	.00849	.00959	.01068

ILLUSTRATION: MATERIAL THICKNESS = .188, INSIDE RADIUS = .375. WHERE THESE TWO MEET (HORIZONTALLY AND VERTICALLY) = .00802. IF THE INSIDE BEND = 115°, THE BEND ALLOWANCE (B/A) = 180° − 115° = 65° .00802 X 65° = .5213″

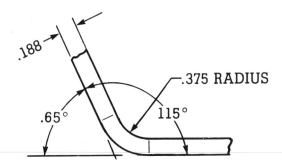

APPENDIX K

SHEET METAL GAGES

NUMBER GAGE	*AMERICAN WIRE GAGE OR BROWN & SHARPE GAGE	**UNITED STATES STANDARD GAGE	MACHINE AND WOOD SCREW GAGE
6/0	.5800	—	—
5/0	.5165	—	—
4/0	.4600	.4063	—
3/0	.4096	.3750	—
2/0	.3648	.3438	—
0	.3249	.3125	.060
1	.2893	.2813	.073
2	.2576	.2656	.086
3	.2294	.2500	.099
4	.2043	.2344	.112
5	.1819	.2188	.125
6	.1620	.2031	.138
7	.1443	.1875	.151
8	.1285	.1719	.164
9	.1144	.1563	.177
10	.1019	.1406	.190
11	.0907	.1250	.203
12	.0808	.1094	.216
13	.0720	.0938	—
14	.0641	.0781	.242
15	.0571	.0703	—
16	.0508	.0625	.268
17	.0453	.0563	—
18	.0403	.0500	.294
19	.0359	.0438	—
20	.0320	.0375	.320
21	.0285	.0344	—
22	.0253	.0313	—
23	.0226	.0281	—
24	.0201	.0250	.372
25	.0179	.0219	—
26	.0159	.0188	—
27	.0142	.0172	—
28	.0126	.0156	—
29	.0113	.0141	—
30	.0100	.0125	.450
31	.0089	.0109	—
32	.0080	.0102	—
33	.0071	.0094	—
34	.0063	.0086	—
35	.0056	.0078	—
36	.0050	.0070	—

* American or Brown & Sharpe Gage: copper wire, brass, copper alloys and nickel silver wire and sheet, also aluminum sheet, rod, and wire.

**United States Standard Gage: steel and Monel metal sheets.
The use of decimals of an inch for dimensions specifying sheet and wire is recommended.

APPENDIX L

AMERICAN NATIONAL STANDARD RUNNING AND SLIDING FITS (ANSI B4.1-1967, R1979)

TOLERANCE LIMITS GIVEN IN BODY OF TABLE ARE ADDED OR SUBTRACTED TO BASIC SIZE (AS INDICATED BY + OR − SIGN) TO OBTAIN MAXIMUM AND MINIMUM SIZES OF MATING PARTS.

	CLASS RC 1			CLASS RC 2			CLASS RC 3			CLASS RC 4		
SIZE RANGE, INCHES	MIN/ MAX CLEAR- ANCE	TOLERANCE LIMITS		CLEAR- ANCE	TOLERANCE LIMITS		CLEAR- ANCE	TOLERANCE LIMITS		CLEAR- ANCE	TOLERANCE LIMITS	
		HOLE	SHAFT		HOLE	SHAFT		HOLD	SHAFT		HOLE	SHAFT
OVER TO	VALUES SHOWN BELOW ARE IN THOUSANDTHS OF AN INCH											
0 - 0.12	.01	+0.2	−0.1	0.1	+0.25	−0.1	0.3	+0.4	−0.3	0.3	+0.6	−0.3
	0.45	0	−0.25	0.55	0	−0.3	0.95	0	−0.55	1.3	0	−0.7
0.12 - 0.24	0.15	+0.2	−0.15	0.15	+0.3	−0.15	0.4	+0.5	−0.4	0.4	+0.7	−0.4
	0.5	0	−0.3	0.65	0	−0.35	1.12	0	−0.7	1.6	0	−0.9
0.24 - 0.40	0.2	+0.25	−0.2	0.2	+0.4	−0.2	0.5	+0.6	−0.5	0.5	+0.9	−0.5
	0.6	0	−0.35	0.85	0	−0.45	1.5	0	−0.9	2.0	0	−1.1
0.40 - 0.71	0.25	+0.3	−0.25	0.25	+0.4	−0.25	0.6	+0.7	−0.6	0.6	+1.0	−0.6
	0.75	0	−0.45	0.95	0	−0.55	1.7	0	−1.0	2.3	0	−1.3
0.71 - 1.19	0.3	+0.4	−0.3	0.3	+0.5	−0.3	0.8	+0.8	−0.8	0.8	+1.2	−0.8
	0.95	0	−0.55	1.2	0	−0.7	2.1	0	−1.3	2.8	0	−1.6
1.19 - 1.97	0.4	+0.4	−0.4	0.4	+0.6	−0.4	1.0	+1.0	−1.0	1.0	+1.6	−1.0
	1.1	0	−0.7	1.4	0	−0.8	2.6	0	−1.6	3.6	0	−2.0
1.97 - 3.15	0.4	+0.5	−0.4	0.4	+0.7	−0.4	1.2	+1.2	−1.2	1.2	+1.8	−1.2
	1.2	0	−0.7	1.6	0	−0.9	3.1	0	−1.9	4.2	0	−2.4

	CLASS RC 5			CLASS RC 6			CLASS RC 7			CLASS RC 8			CLASS RC 9		
SIZE RANGE, INCHES	MIN/ MAX CLEAR- ANCE	TOLERANCE LIMITS		CLEAR- ANCE	TOLERANCE LIMITS		CLEAR- ANCE	TOLERANCE LIMITS		CLEAR- ANCE	TOLERANCE LIMITS		CLEAR- ANCE	TOLERANCE LIMITS	
		HOLE	SHAFT		HOLE	SHAFT		HOLE	SHAFT		HOLE	SHAFT		HOLE	SHAFT
OVER TO	VALUES SHOWN BELOW ARE IN THOUSANDTHS OF AN INCH														
0 - 0.12	0.6	+0.6	−0.6	0.6	+1.0	−0.6	1.0	+1.0	−1.0	2.5	+1.6	−2.5	4.0	+2.5	−4.0
	1.6	0	−1.0	2.2	0	−1.2	2.6	0	−1.6	5.1	0	−3.5	8.1	0	−5.6
0.12 - 0.24	0.8	+0.7	−0.8	0.8	+1.2	−0.8	1.2	+1.2	−1.2	2.8	+1.8	−2.8	4.5	+3.0	−4.5
	2.0	0	−1.3	2.7	0	−1.5	3.1	0	−1.9	5.8	0	−4.0	9.0	0	−6.0
0.24 - 0.40	1.0	+0.9	1.0	1.0	+1.4	−1.0	1.6	+1.4	−1.6	3.0	+2.2	−3.0	5.0	+3.5	−5.0
	2.5	0	−1.6	3.3	0	−1.9	3.9	0	−2.5	6.6	0	−4.4	10.7	0	−7.2
0.40 - 0.71	1.2	+1.0	−1.2	1.2	+1.6	−1.2	2.0	+1.6	−2.0	3.5	+2.8	−3.5	6.0	+4.0	−6.0
	2.9	0	−1.9	3.8	0	−2.2	4.6	0	−3.0	7.9	0	−5.1	12.8	0	−8.8
0.71 - 1.19	1.6	+1.2	−1.6	1.6	+2.0	−1.6	2.5	+2.0	−2.5	4.5	+3.5	−4.5	7.0	+5.9	−7.0
	3.6	0	−2.4	4.8	0	−2.8	5.7	0	−3.7	10.0	0	−6.5	15.5	0	−10.5
1.19 - 1.97	2.0	+1.6	−2.0	2.0	+2.5	−2.0	3.0	+2.5	−3.0	5.0	+4.0	−5.0	8.0	+6.0	−8 0
	4.6	0	−3.0	6.1	0	−3.6	7.1	0	−4.6	11.5	0	−7.5	18.0	0	−12.0
1.97 - 3.15	2.5	+1.8	−2.5	2.5	+3.0	−2.5	4.0	+3.0	−4.0	6.0	+4.5	−6.0	9.0	+7.0	9.0
	5.5	0	−3.7	7.3	0	−4.3	8.8	0	−5.8	13.5	0	−9.0	20.5	0	−13.5

APPENDIX L

ANSI STANDARD FORCE AND SHRINK FITS (ANSI B4.1-1967, R1979)

SIZE RANGE, INCHES OVER – TO	CLASS FN 1 MIN/MAX INTER-FERENCE	TOLERANCE LIMITS HOLE	SHAFT	CLASS FN 2 INTER-FERENCE	TOLERANCE LIMITS HOLE	SHAFT	CLASS FN 3 INTER-FERENCE	TOLERANCE LIMITS HOLE	SHAFT	CLASS FN 4 INTER-FERENCE	TOLERANCE LIMITS HOLE	SHAFT	CLASS FN 5 INTER-FERENCE	TOLERANCE LIMITS HOLE	SHAFT
					VALUES SHOWN BELOW ARE IN THOUSANDTHS OF AN INCH										
0 – 0.12	0.05	+0.25	+0.5	0.2	+0.4	+0.85				0.3	+0.4	+0.95	0.3	+0.6	+1.3
	0.5	0	+0.3	0.85	0	+0.6				0.95	0	+0.7	1.3	0	+0.9
0.12 – 0.24	0.1	+0.3	+0.6	0.2	+0.5	+1.0				0.4	+0.5	+1.2	0.5	+0.7	+1.7
	0.6	0	+0.4	1.0	0	+.07				1.2	0	+0.9	1.7	0	+1.2
0.24 – 0.40	0.1	+0.4	+0.75	0.4	+0.6	+1.4				0.6	+0.6	+1.6	0.5	+0.9	+2.0
	0.75	0	+0.5	1.4	0	+1.0				1.6	0	+1.2	2.0	0	+1.4
0.40 – 0.56	0.1	+0.4	+0.8	0.5	+0.7	+1.6				0.7	+0.7	+1.8	0.6	+1.0	+2.3
	0.8	0	+.05	1.6	0	+1.2				1.8	0	+1.4	2.3	0	+1.6
0.56 – 0.71	0.2	−0.4	−0.9	0.5	+0.7	+1.6				0.7	+0.7	+1.8	0.8	+1.0	+2.5
	0.9	0	−0.6	1.6	0	+1.2				1.8	0	+1.4	2.5	0	+1.8
0.71 – 0.95	0.2	−0.5	−1.1	0.6	+0.8	+1.9				0.8	+0.8	+2.1	1.0	+1.2	+3.0
	1.1	0	−0.7	1.9	0	+1.4				2.1	0	+1.6	3.0	0	+2.2
0.95 – 1.19	0.3	−0.5	−1.2	0.6	−0.8	+1.9	0.8	+0.8	+2.1	1.0	+0.8	+2.3	1.3	+1.2	+3.3
	1.2	0	−08	1.9	0	−1.4	2.1	0	+1.6	2.3	0	+1.8	3.3	0	+2.5
1.19 – 1.58	0.3	−0.6	−1.3	0.8	−1.0	+2.4	1.0	+1.0	+2.6	1.5	+1.0	+3.1	1.4	+1.6	+4.0
	1.3	0	−0.9	2.4	0	+1.8	2.6	0	+2.0	3.1	0	+2.5	4.0	0	+3.0
1.58 – 1.97	0.4	−0.6	−1.4	0.8	−1.0	+2.4	1.2	+1.0	+2.8	1.8	+1.0	+3.4	2.4	+1.6	+5.0
	1.4	0	−1.0	2.4	0	+1.8	2.8	0	+2.2	3.4	0	+2.8	5.0	0	+4.0
1.97 – 2.56	0.6	−0.7	−1.8	0.8	−1.2	+2.7	1.3	+1.2	+3.2	2.3	+1.2	+4.2	3.2	+1.8	+6.2
	1.8	0	−1.3	2.7	0	+2.0	3.2	0	+2.5	4.2	0	+3.5	6.2	0	+5.0
2.56 – 3.15	0.7	−0.7	−1.9	1.0	+1.2	+2.9	1.8	+1.2	+3.7	2.8	+1.2	+4.7	4.2	+1.8	+7.2
	1.9	0	−1.4	2.9	0	+2.2	3.7	0	+3.0	4.7	0	+4.0	7.2	0	+6.0

APPENDIX M

(R) ROTOR CLIP HO RINGS:
Axially Assembled, Internal Retaining Rings

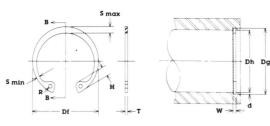

Free Diameter & Ring Measurements
with Section B—B

Housing Diameter &
Groove Dimensions

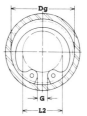

Lug Clearance Diameter & Gap Width

RING NO.	HOUSING Diameter Inches			GROOVE SIZE Diameter		Width		Depth	RING SIZE & WEIGHT Free Diameter		Thickness ‡		Weight Per 1000 pieces	LUG CLEAR. Lugs Compressed in housing	Lugs Released in groove	†THRUST LD. (lbs.) Square corner abutment Ring Safety factor of 4	Groove Safety factor of 2
	Dh DEC	Dh FRACT	Dh MM	Dg	TOL.	W	TOL.	d	Df	TOL.	T	TOL.	LBS.	L1	L2	Pr	Pg
HO-25	.250	1/4	6.4	.268	±.001	.020	+.002	.009	.280		.015		.08	.115	.133	420	190
HO-31	.312	5/16	7.9	.330	.0015*	.020	-.000	.009	.346		.015		.11	.173	.191	530	240
HO-37	.375	3/8	9.5	.397	±.002	.029		.011	.415		.025		.25	.204	.226	1050	350
HO-43	.438	7/16	11.1	.461	.002*	.029		.012	.482		.025		.37	.23	.254	1220	440
HO-45	.453	29/64	11.5	.477		.029		.012	.498		.025		.43	.25	.274	1280	460
HO-50	.500	1/2	12.7	.530		.039		.015	.548	+.010	.035		.70	.26	.29	1980	510
HO-51	.512	—	13.0	.542		.039		.015	.560	-.005	.035		.77	.27	.30	2030	520
HO-56	.562	9/16	14.3	.596	±.002	.039		.017	.620		.035		.86	.275	.305	2220	710
HO-62	.625	5/8	15.9	.665	.004*	.039		.020	.694		.035		1.0	.34	.38	2470	1050
HO-68	.688	11/16	17.5	.732		.039	+.003	.022	.763		.035		1.2	.40	.44	2700	1280
HO-75	.750	3/4	19.0	.796		.039	-.000	.023	.831		.035		1.3	.45	.49	3000	1460
HO-77	.777	—	19.7	.825		.046		.024	.859		.042		1.7	.475	.52	4550	1580
HO-81	.812	13/16	20.6	.862		.046		.025	.901		.042		1.9	.49	.54	4800	1710
HO-86	.866	—	22.0	.920		.046		.027	.961		.042		2.0	.54	.59	5100	1980
HO-87	.875	7/8	22.2	.931	±.003	.046		.028	.971	+.015	.042	±.002	2.1	.545	.60	5150	2080
HO-90	.901	—	22.9	.959	.004*	.046		.029	1.000	-.010	.042		2.2	.565	.62	5350	2200
HO-93	.938	15/16	23.8	1.000		.046		.031	1.041		.042		2.4	.61	.67	5600	2450
HO-100	1.000	1	25.4	1.066		.046		.033	1.111		.042		2.7	.665	.73	5950	2800
HO-102	1.023	—	26.0	1.091		.046		.034	1.136		.042		2.8	.69	.755	6050	3000
HO-106	1.062	1-1/16	27.0	1.130		.056		.034	1.180		.050		3.7	.685	.75	7450	3050
HO-112	1.125	1-1/8	28.6	1.197		.056		.036	1.249		.050		4.0	.745	.815	7900	3400
HO-118	1.181	—	30.0	1.255		.056		.037	1.319		.050		4.3	.79	.86	8400	3700
HO-118	1.188	1-3/16	30.2	1.262		.056		.037	1.319		.050		4.3	.80	.87	8400	3700
HO-125	1.250	1-1/4	31.7	1.330		.056		.040	1.388		.050		4.8	.875	.955	8800	4250
HO-125	1.259	—	32.0	1.339	±.004	.056	+.004	.040	1.388	+.025	.050		4.8	.885	.965	8800	4250
HO-131	1.312	1-5/16	33.3	1.396	.005*	.056	-.000	.042	1.456	-.020	.050		5.0	.93	1.01	9300	4700
HO-137	1.375	1-3/8	34.9	1.461		.056		.043	1.526		.050		5.1	.99	1.07	9700	5050
HO-137	1.378	—	35.0	1.464		.056		.043	1.526		.050		5.1	.99	1.07	9700	5050
HO-143	1.438	1-7/16	36.5	1.528		.056		.045	1.596		.050		5.8	1.06	1.15	10200	5500
HO-145	1.456	—	37.0	1.548		.056		.046	1.616		.050		6.4	1.08	1.17	10300	5700
HO-150	1.500	1-1/2	38.1	1.594		.056		.047	1.660		.050		6.5	1.12	1.21	10550	6000
HO-156	1.562	1-9/16	39.7	1.658		.068		.048	1.734		.062		8.9	1.14	1.23	13700	6350
HO-156	1.575	—	40.0	1.671		.068		.048	1.734		.062		8.9	1.15	1.24	13700	6350
HO-162	1.625	1-5/8	41.3	1.725	±.005	.068		.050	1.804	+.035	.062	±.003	10.0	1.15	1.25	14200	6900
HO-165	1.653	—	42.0	1.755	.005*	.068		.051	1.835	-.025	.062		10.4	1.17	1.27	14500	7200
HO-168	1.688	1-11/16	42.9	1.792		.068		.052	1.874		.062		10.8	1.21	1.31	14800	7450

COURTESY OF ROTOR CLIP CO., INC.

*F.I.M. (Full Indicator Movement) — Maximum allowable deviation of concentricity between groove and housing.
† For an explanation of formulas used to derive thrust load and other performance data, contact the Rotor Clip engineering department.
‡ For plated rings, add .002" to the listed maximum thickness. Maximum ring thickness will be a minimum of .0002" less than the listed groove width (W) minimum.

APPENDIX M

Allowable Corner Radius & Chamfer

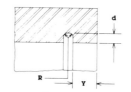

Exploded Groove Profile & Edge Margin (Y)
Maximum bottom radii (R) .005 for ring sizes
–25 thru –100; .010 for ring sizes –102 thru –350.

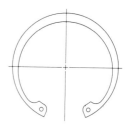

Optional Lug Design
For sizes HO 206 thru HO 350.

RING NO.	LUG HEIGHT		MAXIMUM SECTION		MINIMUM SECTION		HOLE DIAMETER		GAP WIDTH Ring in groove	ALLOWABLE CORNER RADII & CHAMFERS		MAX. LOAD w/ R max or Ch max (in lbs.)	EDGE MARGIN
	H	TOL.	S max	TOL.	S min	TOL.	R	TOL.	G	R max	Ch max	P'r	Y
HO-25	.065		.025		.015		.031		.047	.011	.0085	190	.027
HO-31	.066	±.003	.033	±.002	.018	±.002	.031		.055	.016	.013	190	.027
HO-37	.082		.040		.028		.041		.063	.023	.018	530	.033
HO-43	.098		.049	±.003	.029	±.003	.041		.063	.027	.021	530	.036
HO-45	.098		.050		.030		.047		.071	.027	.021	530	.036
HO-50	.114		.053		.035		.047		.090	.027	.021	1100	.045
HO-51	.114		.053		.035		.047		.092	.027	.021	1100	.045
HO-56	.132		.053		.035		.047		.095	.027	.021	1100	.051
HO-62	.132		.060	±.004	.035	±.004	.062		.104	.027	.021	1100	.060
HO-68	.132		.063		.036		.062	+.010 −.002	.118	.027	.021	1100	.066
HO-75	.142		.070		.040		.062		.143	.032	.025	1100	.069
HO-77	.146		.074		.044		.062		.145	.035	.028	1650	.072
HO-81	.155		.077		.044		.062		.153	.035	.028	1650	.075
HO-86	.155		.081		.045		.062		.172	.035	.028	1650	.081
HO-87	.155		.084	±.005	.045	±.005	.062		.179	.035	.028	1650	.084
HO-90	.155		.087		.047		.062		.188	.038	.030	1650	.087
HO-93	.155		.091		.050		.062		.200	.038	.030	1650	.093
HO-100	.155		.104		.052		.062		.212	.042	.034	1650	.099
HO-102	.155		.106		.054		.062		.220	.042	.034	1650	.102
HO-106	.180		.110		.055		.078		.213	.044	.035	2400	.102
HO-112	.180		.116		.057		.078		.232	.047	.036	2400	.108
HO-118	.180		.120		.058		.078		.226	.047	.036	2400	.111
HO-118	.180	±.005	.120		.058		.078		.245	.047	.036	2400	.111
HO-125	.180		.124		.062		.078		.265	.048	.038	2400	.120
HO-125	.180		.124	±.006	.062	±.006	.078		.290	.048	.038	2400	.120
HO-131	.180		.130		.062		.078		.284	.048	.038	2400	.126
HO-137	.180		.130		.063		.078	+.015 −.002	.297	.048	.038	2400	.129
HO-137	.180		.130		.063		.078		.305	.048	.038	2400	.129
HO-143	.180		.133		.065		.078		.313	.048	.038	2400	.135
HO-145	.180		.133		.065		.078		.320	.048	.038	2400	.138
HO-150	.180		.133		.066		.078		.340	.048	.038	2400	.141
HO-156	.202		.157		.078		.078		.338	.064	.050	3900	.144
HO-156	.202		.157		.078		.078		.374	.064	.050	3900	.144
HO-162	.220		.164	±.007	.082	±.007	.078		.339	.064	.050	3900	.150
HO-165	.227		.167		.083		.078		.348	.064	.050	3900	.153
HO-168	.227		.170		.085		.078		.357	.064	.050	3900	.156

COURTESY OF ROTOR CLIP CO., INC.

APPENDIX M

ⓡ ROTOR CLIP SH RINGS:
Axially Assembled, External Retaining Rings

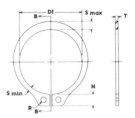

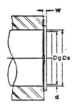

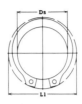

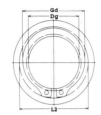

Free Diameter & Ring Measurements with Section B—B	Shaft Diameter & Groove Dimensions	Lug Clearance Diameter & Gaging Diameter

Column groups: SHAFT — Diameter (Inches): Ds DEC, Ds FRACT, Ds MM · GROOVE SIZE — Diameter (Dg, TOL.), Width (W, TOL.), Depth (d) · RING SIZE & WEIGHT — Free Diameter (Df, TOL.), Thickness ‡ (T, TOL.), Weight Per 1000 pieces (LBS.) · LUG CLEAR. — Lugs Expanded over shaft (L1), Lugs Released in groove (L2) · †THRUST LD. (lbs.) Square corner abutment — Ring Safety factor of 4 (Pr), Groove Safety factor of 2 (Pg)

RING NO.	Ds DEC	Ds FRACT	Ds MM	Dg	TOL.	W	TOL.	d	Df	TOL.	T	TOL.	LBS.	L1	L2	Pr	Pg
**SH-12	125	1/8	3.2	.117		.012		.004	.112		.010	±.001	.018	.222	.214	110	35
**SH-15	156	5/32	4.0	.146		.012		.005	.142		.010		.037	.270	.260	130	55
**SH-18	188	3/16	4.8	.175	±.0015	.018	+.002	.006	.168	+.002	.015		.059	.298	.286	240	80
**SH-19	197	—	5.0	.185	.0015*	.018	−.000	.006	.179	−.004	.015		.063	.319	.307	250	85
**SH-21	219	7/32	5.6	.205		.018		.007	.196		.015		.074	.338	.324	280	110
**SH-23	236	15/64	6.0	.222		.018		.007	.215		.015		.086	.355	.341	310	120
SH-25	250	1/4	6.4	.230		.029		.010	.225		.025		.21	.45	.43	590	175
SH-27	276	—	7.0	.255		.029		.010	.250		.025		.23	.48	.46	650	195
SH-28	281	9/32	7.1	.261		.029		.010	.256		.025		.24	.49	.47	660	200
SH-31	312	5/16	7.9	.290		.029		.011	.281		.025		.27	.54	.52	740	240
SH-34	344	11/32	8.7	.321		.029		.011	.309		.025		.31	.57	.55	800	265
SH-35	354	—	9.0	.330	±.002	.029		.012	.320	+.002	.025		.35	.59	.57	820	300
SH-37	375	3/8	9.5	.352	.002*	.029		.012	.338	−.005	.025		.39	.61	.59	870	320
SH-39	394	—	10.0	.369		.029		.012	.354		.025		.42	.62	.60	940	335
SH-40	406	13/32	10.3	.382		.029		.012	.366		.025		.43	.63	.61	950	350
SH-43	438	7/16	11.1	.412		.029		.013	.395		.025		.50	.66	.64	1020	400
SH-46	469	15/32	11.9	.443		.029		.013	.428		.025		.54	.68	.66	1100	450
SH-50	500	1/2	12.7	.468	±.002	.039		.016	.461		.035		.91	.77	.74	1650	550
SH-55	551	—	14.0	.519	.004*	.039		.016	.509		.035		.90	.81	.78	1800	600
SH-56	562	9/16	14.3	.530		.039	+.003	.016	.521		.035	±.002	1.1	.82	.79	1850	650
SH-59	594	19/32	15.1	.559		.039	−.000	.017	.550		.035		1.2	.86	.83	1950	750
SH-62	625	5/8	15.9	.588		.039		.018	.579		.035		1.3	.90	.87	2060	800
SH-66	669	—	17.0	.629		.039		.020	.621		.035		1.4	.93	.89	2200	950
SH-66	672	43/64	17.1	.631		.039		.020	.621		.035		1.4	.93	.89	2200	950
SH-68	688	11/16	17.5	.646		.046		.021	.635		.042		1.8	1.01	.97	3400	1000
SH-75	750	3/4	19.0	.704		.046		.023	.693	+.005	.042		2.1	1.09	1.05	3700	1200
SH-78	781	25/32	19.8	.733	±.003	.046		.024	.722	−.010	.042		2.3	1.12	1.08	3900	1300
SH-81	812	13/16	20.6	.762	.004*	.046		.025	.751		.042		2.5	1.15	1.10	4000	1450
SH-87	875	7/8	22.2	.821		.046		.027	.810		.042		2.8	1.21	1.16	4300	1650
SH-93	938	15/16	23.8	.882		.046		.028	.867		.042		3.1	1.34	1.29	4650	1850
SH-98	984	63/64	25.0	.926		.046		.029	.910		.042		3.5	1.39	1.34	4850	2000
SH-100	1.000	1	25.4	.940		.046		.030	.925		.042		3.6	1.41	1.35	4950	2100
SH-102	1.023	—	26.0	.961		.046		.031	.946		.042		3.9	1.43	1.37	5050	2250
SH-106	1.062	1-1/16	27.0	.998	±.004	.056	+.004	.032	.982	+.010	.050		4.8	1.50	1.44	6200	2400
SH-112	1.125	1-1/8	28.6	1.059	.005*	.056	−.000	.033	1.041	−.015	.050		5.1	1.55	1.49	6600	2600

COURTESY OF ROTOR CLIP CO., INC.

*F.I.M. (Full Indicator Movement) — Maximum allowable deviation of concentricity between groove and shaft.

† For an explanation of formulas used to derive thrust load and other performance data, contact the Rotor Clip engineering department.

‡ For plated rings, add .002" to the listed maximum thickness. Maximum ring thickness will be a minimum of .0002" less than the listed groove width (W) minimum.

** Sizes -12 through -23 available in Beryllium Copper only.

APPENDIX M

Maximum Corner Radius & Chamfer

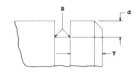

**Exploded Groove Profile &
Edge Margin (Y)**
Maximum bottom radii (R), sharp corners for
ring sizes –12 thru –23. .003 for sizes –25 thru
–35. .005 for sizes –37 thru –100. .010 for sizes
–102 thru –350.

Lug Design
For sizes SH-12 thru SH-23

RING NO.	LUG HEIGHT		MAXIMUM SECTION		MINIMUM SECTION		HOLE DIAMETER		GAG-ING DIA.	ALLOWABLE CORNER RADII & CHAMPERS		MAX. LOAD w/R max or Ch max (in lbs.)	EDGE MAR-GIN	R.P.M. LIMITS Standard material
	H	TOL.	S max	TOL.	S min	TOL.	R	TOL.	Gd	R max	Ch max	P'r	Y	
SH-12	.046		.018	±.0015	.011	±.0015	.026		.148	.010	.006	45	.012	80000
SH-15	.054		.026		.016		.026		.189	.015	.009	45	.015	80000
SH-18	.050	±.002	.025		.016		.025		.218	.014	.0085	105	.018	80000
SH-19	.056		.026	±.002	.016	±.002	.026		.229	.0145	.009	105	.018	80000
SH-21	.056		.028		.017		.026		.252	.015	.009	105	.021	80000
SH-23	.056		.030		.019		.026		.272	.0165	.010	105	.021	80000
SH-25	.080		.035		.025		.041		.290	.018	.011	470	.030	80000
SH-27	.081		.035		.024		.041		.315	.0175	.0105	470	.031	76000
SH-28	.080		.038		.0255		.041		.326	.020	.012	470	.030	74000
SH-31	.087		.040		.026		.041		.357	.020	.012	470	.033	70000
SH-34	.087		.042		.0265		.041		.390	.021	.0125	470	.033	64000
SH-35	.087		.046	±.003	.029	±.003	.041		.405	.023	.014	470	.036	62000
SH-37	.088		.050		.0305		.041		.433	.026	.0155	470	.036	60000
SH-39	.087		.052		.031		.041		.452	.027	.016	470	.037	56500
SH-40	.087	±.003	.054		.033		.041	+.010 −.002	.468	.0285	.017	470	.036	55000
SH-43	.088		.055		.033		.041		.501	.029	.0175	470	.039	50000
SH-46	.088		.060		.035		.041		.540	.031	.018	470	.039	42000
SH-50	.108		.065		.040		.047		.574	.034	.020	910	.048	40000
SH-55	.108		.053		.036		.047		.611	.027	.0165	910	.048	36000
SH-56	.108		.072		.041		.047		.644	.038	.023	910	.048	35000
SH-59	.109		.076	±.004	.043	±.004	.047		.680	.0395	.0235	910	.052	32000
SH-62	.110		.080		.045		.047		.715	.0415	.025	910	.055	30000
SH-66	.110		.082		.043		.047		.756	.040	.024	910	.060	29000
SH-66	.110		.082		.043		.047		.758	.040	.024	910	.060	29000
SH-68	.136		.084		.048		.052		.779	.042	.025	1340	.063	28000
SH-75	.136		.092		.051		.052		.850	.046	.0275	1340	.069	26500
SH-78	.136		.094		.052		.052		.883	.047	.028	1340	.072	25500
SH-81	.136		.096		.054		.052		.914	.047	.028	1340	.075	24500
SH-87	.137		.104	±.005	.057	±.005	.052		.987	.051	.0305	1340	.081	23000
SH-93	.166	±.004	.110		.063		.078		1.054	.055	.033	1340	.084	21500
SH-98	.167		.114		.0645		.078	+.015 −.002	1.106	.056	.0335	1340	.087	20500
SH-100	.167		.116		.065		.078		1.122	.057	.034	1340	.090	20000
SH-102	.168		.118		.066		.078		1.147	.058	.035	1340	.093	19500
SH-106	.181		.122	±.006	.069	±.006	.078		1.192	.060	.036	1950	.096	19000
SH-112	.182		.128		.071		.078		1.261	.063	.038	1950	.099	18800

APPENDIX N

KEYS

KEYS—SQUARE, FLAT, PLAIN TAPER, AND GIB HEAD

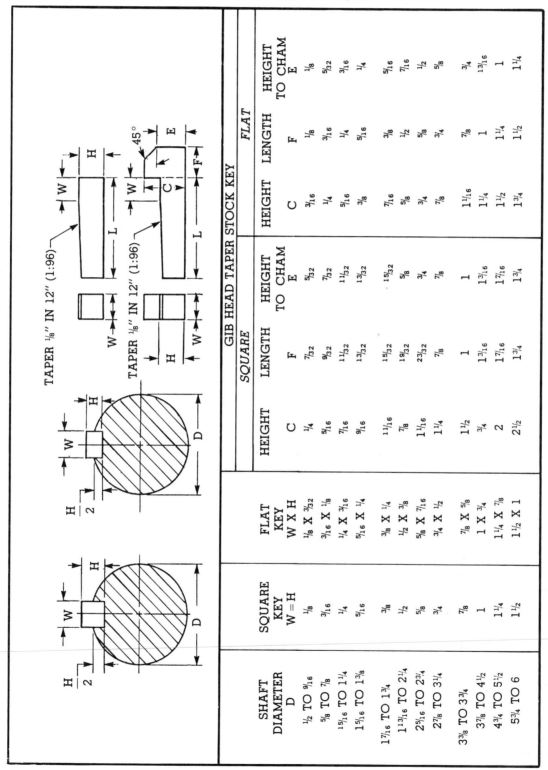

TAPER ⅛″ IN 12″ (1:96)

SHAFT DIAMETER D	SQUARE KEY W = H	FLAT KEY W X H	GIB HEAD TAPER STOCK KEY — SQUARE			GIB HEAD TAPER STOCK KEY — FLAT		
			HEIGHT C	LENGTH F	HEIGHT TO CHAM E	HEIGHT C	LENGTH F	HEIGHT TO CHAM E
½ TO 9/16	⅛	⅛ X 3/32	¼	7/32	5/32	3/16	⅛	⅛
5/8 TO 7/8	3/16	3/16 X ⅛	5/16	9/32	7/32	¼	3/16	5/32
15/16 TO 1¼	¼	¼ X 3/16	7/16	11/32	11/32	5/16	¼	3/16
1 5/16 TO 1 3/8	5/16	5/16 X ¼	9/16	13/32	13/32	3/8	5/16	¼
1 7/16 TO 1¾	3/8	3/8 X ¼	11/16	15/32	15/32	7/16	3/8	5/16
1 13/16 TO 2¼	½	½ X 3/8	7/8	19/32	5/8	5/8	½	7/16
2 5/16 TO 2¾	5/8	5/8 X 7/16	1 1/16	23/32	¾	¾	5/8	½
2 7/8 TO 3¼	¾	¾ X ½	1¼	7/8	7/8	7/8	¾	5/8
3 3/8 TO 3¾	7/8	7/8 X 5/8	1½	1	1	1 1/16	7/8	¾
3 7/8 TO 4½	1	1 X ¾	1¾	1 3/16	1 3/16	1¼	1	13/16
4¾ TO 5½	1¼	1¼ X 7/8	2	1 7/16	1 7/16	1½	1¼	1
5¾ TO 6	1½	1½ X 1	2½	1¾	1¾	1¾	1½	1¼

APPENDIX N

KEYS

WOODRUFF KEYS—AMERICAN NATIONAL STANDARD

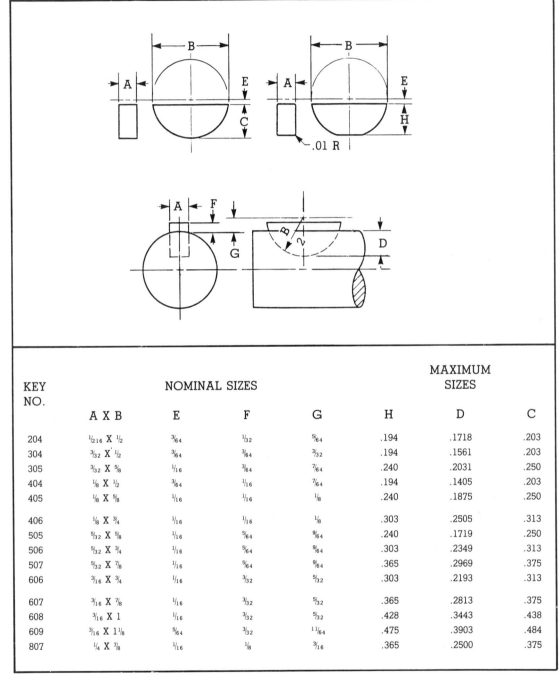

KEY NO.	NOMINAL SIZES					MAXIMUM SIZES	
	A X B	E	F	G	H	D	C
204	$\frac{1}{216}$ X $\frac{1}{2}$	$\frac{3}{64}$	$\frac{1}{32}$	$\frac{5}{64}$	.194	.1718	.203
304	$\frac{3}{32}$ X $\frac{1}{2}$	$\frac{3}{64}$	$\frac{3}{64}$	$\frac{3}{32}$	.194	.1561	.203
305	$\frac{3}{32}$ X $\frac{5}{8}$	$\frac{1}{16}$	$\frac{3}{64}$	$\frac{7}{64}$	.240	.2031	.250
404	$\frac{1}{8}$ X $\frac{1}{2}$	$\frac{3}{64}$	$\frac{1}{16}$	$\frac{7}{64}$	.194	.1405	.203
405	$\frac{1}{8}$ X $\frac{5}{8}$	$\frac{1}{16}$	$\frac{1}{16}$	$\frac{1}{8}$	.240	.1875	.250
406	$\frac{1}{8}$ X $\frac{3}{4}$	$\frac{1}{16}$	$\frac{1}{16}$	$\frac{1}{8}$	.303	.2505	.313
505	$\frac{5}{32}$ X $\frac{5}{8}$	$\frac{1}{16}$	$\frac{5}{64}$	$\frac{9}{64}$	.240	.1719	.250
506	$\frac{5}{32}$ X $\frac{3}{4}$	$\frac{1}{16}$	$\frac{5}{64}$	$\frac{9}{64}$	.303	.2349	.313
507	$\frac{5}{32}$ X $\frac{7}{8}$	$\frac{1}{16}$	$\frac{5}{64}$	$\frac{9}{64}$	.365	.2969	.375
606	$\frac{3}{16}$ X $\frac{3}{4}$	$\frac{1}{16}$	$\frac{3}{32}$	$\frac{5}{32}$	.303	.2193	.313
607	$\frac{3}{16}$ X $\frac{7}{8}$	$\frac{1}{16}$	$\frac{3}{32}$	$\frac{5}{32}$	.365	.2813	.375
608	$\frac{3}{16}$ X 1	$\frac{1}{16}$	$\frac{3}{32}$	$\frac{5}{32}$	.428	.3443	.438
609	$\frac{3}{16}$ X $1\frac{1}{8}$	$\frac{5}{64}$	$\frac{3}{32}$	$\frac{11}{64}$	.475	.3903	.484
807	$\frac{1}{4}$ X $\frac{7}{8}$	$\frac{1}{16}$	$\frac{1}{8}$	$\frac{3}{16}$	.365	.2500	.375

ANSI B 17.2 - 1967 (R1978)

APPENDIX N

KEYS

PRATT AND WHITNEY ROUND-END KEYS

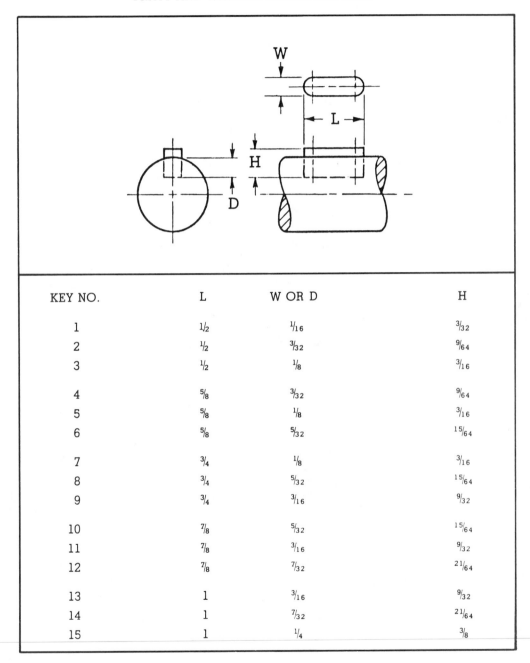

KEY NO.	L	W OR D	H
1	$1/2$	$1/16$	$3/32$
2	$1/2$	$3/32$	$9/64$
3	$1/2$	$1/8$	$3/16$
4	$5/8$	$3/32$	$9/64$
5	$5/8$	$1/8$	$3/16$
6	$5/8$	$5/32$	$15/64$
7	$3/4$	$1/8$	$3/16$
8	$3/4$	$5/32$	$15/64$
9	$3/4$	$3/16$	$9/32$
10	$7/8$	$5/32$	$15/64$
11	$7/8$	$3/16$	$9/32$
12	$7/8$	$7/32$	$21/64$
13	1	$3/16$	$9/32$
14	1	$7/32$	$21/64$
15	1	$1/4$	$3/8$

Glossary

Acute angle An angle of less than 90 degrees.

Allowance The tightest fit between two mating parts. There are two types of allowance: interference and clearance.

Alpha characters The 26 letters of the alphabet (A through Z).

ANSI American National Standards Institute. A professional organization that proposes, develops, modifies, approves, and publishes drafting and manufacturing standards in the United States.

Arc A small segment of a circle or a curve.

Assembly drawing A drawing depicting a product in its completed form. It is a combination of detail parts which are joined to form a subassembly or a complete unit.

Automated drafting station A graphics workstation for producing computer-aided drafting (CAD) data, consisting of two basic elements, hardware and software.

Auxiliary view A view that illustrates the true shape and form of a surface.

Backup disk A disk containing a copy of another disk, created for protection in case one disk becomes damaged or lost.

Basic dimension A numerical value used to describe the theoretical exact size, profile, orientation, or location of a feature. It is the basis on which permissible variations are established.

Baud Data transmission rate of one bit per second.

Bend allowance A length of material around a bend from bend line to bend line.

Bend angle The angle to which sheet metal is bent. It is measured from the flat through the bend to the finished angle after bending.

Bend radius A minimum radius to which material can be bent without fracturing or cracking at the outside radius.

Binary The base 2 numbering system, which uses only the digits 0 and 1.

Bisect To divide in half.

Bit Taken from "binary digit": a 0 or 1 signal.

CAD Computer-aided drafting.

Chamfer An angle that may be produced internally or externally to eliminate sharp edges or burrs on machined objects.

Chassis A thin-gage metal structure or base that is designed to support electronic, electromechanical, or mechanical devices. It may or may not have a cover or be enclosed.

Chip A small segment of silicon material containing electronic devices which perform functions within a computer.

Circle A closed plane curve of which all points are equidistant from the center.

Command An instruction to the computer system to perform a function.

Compass An instrument used to draw circles and arcs.

Computer A popular name referring to the central processing unit (CPU).

Cone A conical-shape object.

Counterbored hole A machined area which provides a flat, flush seat for a fastener.

Countersink or counterdrill A flush or below-surface mounting condition for a screw or a rivet.

Crosshatching Lines that show clearly which surface has been theoretically cut and assist the observer to better understand the internal shape of the object. Also referred to as **section lining**.

Curve A tool used to draw irregular or noncircular shapes which are frequently required in producing development drawings and ellipse-, parabola-, and hyperbola-type drawings.

Datum A theoretically exact point, axis, or plane. A datum is the origin from which the locations or geometric characteristics of features of a part are established.

Descriptive geometry The practice of projecting three-dimensional objects onto a two-dimensional medium so as to solve problems involving lines, angles, and shapes.

Design layout drawing   A drawing that graphically illustrates a design concept and is the origin or basis for all detail and assembly drawings.

Detail drawing   A drawing that presents all the necessary information to fabricate an item, including shape description, size description, and specifications of the part.

Development drawing   Usually a design layout of a pattern or a template. It is drawn and developed in a single plane by the drafter in preparation for folding, rolling, or bending to form some predetermined shape.

Diameter   The length of a straight line running through the center of a circle.

Diazo   A system of reproducing engineering drawings. The diazo process is based on the decomposition of light-sensitive paper into a colorless substance when exposed to ultraviolet light.

Digitizing   "Touching down" a stylus or puck at a particular location; the method by which data is entered on a graphics tablet.

Dimension   A numerical value expressed in appropriate units of measure. Dimensions are used to complete the description of an object or a part.

Dimension line   A line with terminations called arrowheads that shows the direction and extent of a dimension.

Diskette   Another name for a **disk** or **floppy disk**.

Display   The screen on which a drawing is displayed; a method of representing information in visible form.

Dividers   A tool used to transfer measurements, divide lines, or lay out a series of equal distances required in geometric constructions, in descriptive geometry, or on development drawings.

Dot matrix   A group of closely spaced dots with a printed pattern which looks like the shapes of the desired character.

DOD   United States Department of Defense.

Down time   Any interval of time when the CAD system is not available or not working.

Drafting   A method of communication and the language of industry. It is the art of mechanical drawing as it pertains to the field of engineering.

Drawing field   The main body of the medium which is the area to be used for drawing all the necessary views, including dimensions and text, to complete the drawing.

Drawing medium   A material (paper, cloth, or film base) for producing technical drawings.

Dual-dimensioning system   A system of measurement in which units of both the ISO (metric) and inch-foot systems are used simultaneously.

Electric eraser   A device that provides a rapid method of making corrections or changes. More and more drafters are using the electric eraser, which is available in both cord and cordless types.

Ellipse   A closed symmetrical curve that resembles a flattened circle.

Enter   To put information into a computer system.

Eraser   A manual device used to make corrections or changes to technical drawings.

Erasing shield   An ultra-thin piece of metal, with cutouts of various shapes, used as an aid in erasing segments of lines, arcs, and circles.

Execute   To carry out a command or a series of commands.

Extension line   A line, also referred to as a **projection line**, that is used to indicate the extension of a surface or point to a location outside the part outline.

Fasteners   Hardware items, both threaded and unthreaded, that provide an important, fundamental method for joining parts together. There are three basic categories of fasteners: the removable type, the semipermanent type, and the permanent type.

Feature   A general term which refers to a physical attribute of a part, such as a surface, hole, or slot.

File   A collection of information treated as a unit, with an identifying name assigned to it.

Flat washer   A flat, ring-shape device used to distribute the fastening load over an area larger than that of the head of a screw or a nut.

Floppy disk   A portable, flat, black or brown circular plate of thin vinyl with a magnetic coating that is inserted into and removed from the floppy disk drive. It is a software package for storing data, housed in standard square sizes of 3½, 5¼, or 8 inches.

Font   A complete assortment of alphanumeric characters and symbols of a particular size and style which can be used for the several different applications required by the drafter. Font means style.

Function   A type of calculation performed by a computer system; examples are geometric functions and text functions.

Geometric construction   A method of accurately developing geometric shapes ranging from squares, triangles, and three-dimensional cylinders to complex irregular curves and ellipses.

Geometric tolerancing   A method of tolerancing that provides greater quality assurance through the use of symbols and other geometric characteristics.

Grip length The sum of any combination of material thickness plus the thickness of any hardware under the screw head.

Hard copy A drawing produced by a hard-copy unit.

Hard disk A nonremovable disk with a large capacity for storing bits of information in the form of files. Sometimes referred to as a **Winchester disk**.

Hardware The physical equipment necessary for CAD system operation.

Hexagon A polygon having six angles and six sides.

Hieroglyphics A graphic language used by the Egyptians between 4000 and 3000 B.C.

Host One central place where CAD data resides.

IC Integrated circuit. Same as a **chip**.

Input To place data into the information-processing system of a computer.

ISO The International Organization for Standardization, which was formed to develop a single international system of metric screw threads.

Isometric drawing A drawing shown on a single plane or flat surface, but in reality a three-dimensional drawing because height, width, and depth are all pictorially described.

Keyseat A slot in a shaft or a hub, or both, to hold a key.

Lead A writing material used to produce technical drawings. Graphite leads are used on all paper surfaces while plastic leads are normally used on polyester film media.

Lead holder A device used for holding leads for the purpose of drawing.

Lead pointer A lead sharpener used by drafters. It produces the points on leads used in lead holders. The pointer produces a conical-shape point that allows the drafter to produce quality lines on drawings.

Leader A line which is used to direct a dimension, note, or symbol to the intended location on a drawing.

Least Material Condition (LMC) A condition wherein a feature of size contains the least amount of material within the stated limits of size.

Lettering aids Devices that help ensure that freehand lettering on a technical drawing is of good quality by providing uniformity in height and width of characters. An example of a commercially available template for drawing guide lines is the **Ames Lettering Guide**.

Limits or limit dimensions The largest and smallest acceptable dimensions.

Line convention A system of line weights and appearances, developed by the American National Standards Institute (ANSI), consisting of a set of symbols used for representing or describing various parts of an object.

Mainframe A CPU with a large capacity and many terminals for multipurpose use.

Manual drafting equipment Items such as drafting tables, drafting machines, reference tables, and blueline print machines used to produce drawings through mechanical means.

Maximum Material Condition (MMC) A condition wherein a feature of size contains the maximum amount of material within the stated limits of size.

Measurement The determination or estimation of values, dimensions, or quantities by comparison with some known standard. There are several systems of measurement that are used throughout the world, including the metric system, the inch-foot system, the dual-dimensioning system, and the redefined metric system (SI system).

Menu An area of the digitizer containing preassigned commands that control computer system operations.

Message Any combination of words or sentences displayed on the screen that inform the user about commands being executed or information needed by the system.

Meter The unit of distance on which the metric system is based. A meter is equal to 39.37 inches. A meter is divided into 100 units called centimeters (cm), and each centimeter is divided into ten units called millimeters (mm).

Microcomputer The smallest type of CAD system. Microcomputers are dedicated units using personal computers or can be "networked" to host or mainframe CPU's.

Microprocessor The central processing unit (CPU) of a microcomputer.

Minicomputer A CAD system with capabilities ranging between those of a microcomputer and a mainframe. Because of their size, they are often referred to as "desk-top" computers. This type of CAD system is commonly used in small-to-medium-size industries.

Miter line A type of line that provides a rapid and accurate method of projecting point, line, or surface measurements from one view to another.

Monitor The unit that displays an image or drawing; same as **CRT, graphics display**, or **screen**.

Nut A fastener with internal threads that is used in conjunction with a screw or a bolt for attaching parts together.

Oblique drawing A drawing consisting of three axes: the vertical, the horizontal, and the receding. The receding axis may be drawn at an angle 15 to 60 degrees to the horizontal axis. The front view is the most significant view and is drawn to its true scale and shape.

Ogee curve A type of curve used to connect two parallel lines. It is also referred to as a **reverse S-curve**.

On-the-tube A term used to describe an operator who is working at a computer terminal.

Operator A general term for a CAD user.

Output The result of input. It is data that the computer system produces.

Pan To move the display of a drawing on a screen, much as a camera is moved to find the best picture.

Parts list A list of materials identifying those items which are required to complete an assembly or all the necessary parts of some functional grouping.

Pentagon A polygon having five angles and five sides.

Peripheral Additional equipment used in conjunction with, but not as part of, a computer system.

Perpendicular The meeting of a given line or surface at exactly right angles (90 degrees) with another line or surface.

Perspective drawing A projection technique for representing three-dimensional objects and depth relationships on a flat plane or surface.

Pictorial drawing A drawing that portrays what the eye sees when viewing an object and that shows all three dimensions of an object in a single view.

Plot The final drawing as it is drawn by the plotter on paper, vellum, or polyester film.

Polygon A closed plane figure having three or more straight sides.

Precedence of lines A preferred standard order of lines on a drawing—that is, which lines hold preference or precedence over others.

Principal view The view that most completely shows an object's characteristic shape.

Quadrilateral A polygon having four straight sides.

Rack A thin-gage metal structure that includes long vertical members. It is normally made of steel or aluminum and houses electronic, electromechanical, or mechanical equipment.

Radius The length of a straight line running from the center of a circle to the perimeter of the circle.

Raster A network or matrix of dots, each of which falls within a square area known as a **pixel**.

Reference dimension A dimension, usually without tolerance, used for information purposes only.

Remote station A remotely located workstation or system connected to the local system with a modem to enable exchange of information over telephone lines.

Resolution The number of addressable dots per unit area. Low-resolution screens produce jagged, stepped lines.

Retaining ring A precision-type fastener that is used to secure components to shafts or limit the movement of parts in an assembly. Retaining rings are produced in two basic types: external and internal.

Rib A part or piece which serves to shape, support, or strengthen an object.

Rivet A rod with some style of head at one end for permanent fastening applications. Rivets are often used for fastening steel or aluminum sheets for boats or aircraft structures.

Scale A device made of wood, plastic, or metal that is used for determining measurements as well as for showing parts at a reduced or enlarged, suitable size.

Screen The display on a video monitor on which drawings or messages are viewed.

Screw thread A spiral-type groove around a bar, rod, or cylinder for fastening two or more parts together.

Sectional view A view used to clearly show the internal form and detail of an object. The sectional view is obtained by passing an imaginary line through some specific part of the object and drawing the object as though the cut-off part had been removed.

Slot An elongated hole with curved or square ends.

Software The collection of programs used to control the internal functioning and hardware of a computer system.

Spoke A rod, finger, or brace that connects the hub and rim of a wheel.

Spotface To provide an accurately machined area for the head of a fastener or a washer.

Spring lock washer A washer used to place an axial load between a bolt and a nut so that vibration will not easily unthread the fastening.

Stylus A manually operated device that provides input to a display unit.

Surface The exterior or face of a three-dimensional object.

Tangent A line or curve that touches (without crossing) a single point of a circle or an arc.

Technical sketch A drawing used for communicating a concept, idea, or plan in a rapid but effective manner. It is always produced freehand.

Template A thin (usually 1/32 to 1/16 inch thick) plastic aid for producing technical drawings. It is a time-saving device that allows rapid and accurate drawing of shapes and forms which are difficult to produce by other methods.

Terminal A popular name for the combination of a visual display screen (CRT) and keyboard, sometimes referred to as a video display terminal (VDT).

Text On technical drawings, text consists of dimensions, notes, charts, instructions, specifications, legends, lists of materials, and any other information that is best communicated through the use of alphanumeric characters.

Third-angle projection A drawing showing different views of an object (front, top, and side) systematically arranged on a drawing medium to convey necessary information to the reader. The projection plane is observed to be between the viewer and the object, with the views being projected forward to that plane. In **first-angle projection**, the projection plane is observed to be on the far side of the object from the viewer.

Title block The area of a drawing that contains pertinent information that is not provided directly on the drawing field itself.

Tolerance The total amount by which a specific dimension is permitted to vary. The tolerance is the difference between the maximum and minimum limits.

Tools Important aids in making certain that clear, detailed, and concise graphics meet acceptable industrial standards.

Transition piece A piece utilized whenever it is necessary to connect openings in two parts having different geometric shapes.

Triangle A plastic tool used for producing 30, 60, 45, and 90-degree angles (fixed triangles). An adjustable triangle is capable of producing any desired angle.

True position The exact location of a feature established by basic dimensions.

Truncate To cut off a geometric solid at an angle.

Turnkey A system that operates with a variety of software and whose hardware is functional, complete, and ready to use as soon as installation is complete (ready at the turn of a key).

User Same as **operator**.

Web An interior piece of metal plate or sheet which is used to connect heavier sections of an object.

Winchester disk A type of hard disk that is non-removable.

Working drawing A form of technical drawing that presents information, ideas, and instructions in pictorial and text form. Working drawings are usually classified into two groups: detail drawings and assembly drawings.

Index